SNATCHES
OF
BRILLIANCE

SNATCHES
OF
BRILLIANCE

**Words of love, light and extreme wisdom
for the sole purpose of teaching hope, love
and peace, emanating from God's words**

Author:
Carole M. Day

Grosvenor House
Publishing Limited

This book is published by
Grosvenor House Publishing Ltd
Link House
140 The Broadway, Tolworth, Surrey, KT6 7HT.
www.grosvenorhousepublishing.co.uk

A CIP record for this book
is available from the British Library

ISBN 978-1-78623-319-6

To My God

This book is dedicated to my God and Divine Father, to Archangel Michael my Guardian Angel and all the dear souls of the Spirit World wishing for their beautiful words and prayers to be included in this book.

Thank you my dear Father for giving to me this awesome challenge, allowing me to receive these words of love, light and extreme wisdom for the sole purpose of teaching hope, love and peace, emanating from God's words.

~✳~

CONTENTS

Foreword

~✳~

From Our Father Himself

Whosoever Reads This Splendid Book

God our Father in Heaven has asked that whosoever reads this splendid book does so with their hearts open, willing to take in and digest the majestic, loving, Divine messages.

We can with our thoughts and special deliberations save our beautiful planet. God is awesome, a wonder to behold. He wishes with all his prayers that earthly soul's think about, wonder about, and deliberate over, and then know, all that is told in truth in these words, to be all being and truth. Believe whole heartedly what He is telling you in truth.

The world is nigh and tempestuous but its beauty is for you all to see and to experience. Look at what is all around you, such colour and sounds, even below the water where there is much beauty yet unseen.

Pray to me, your Lord and Master, speak to me about daily happenings, however non important it would seem to you. Although your life's venture given to you in a plan before you were sent to the earth's plain cannot be altered, but with your forethought and spiritual strength and asking in your prayers to Me the chance to progress your spiritualism in your time on earth could be successful if you believe.

God Be With You Always

Words by Archangel Michael
10th February 2011

First Divine Message For This Book

Devoted to our Lord on High, Divine prayers are told to all who are open to His authority.

Jesus, Son of God our Father, has asked to be reminded to all souls on earth who think of Him and those about Him.

Wonders to behold are earthly barriers to beauty, solitude and Divine tracks of time, battling with all dimensions of our universe. In this universe, our Heaven and Spirit World, where all live in perfect harmony and multitudes of thoughts leading to miracles and prayers of the highest order battling with hopes, glorification, diversification and all wondrous hopes of glory.

Our lives here in Heaven are satisfying, giving out the love of our very Lord on High. He gives out to all His glory, His love, but uppermost His greatness with kindness giving to us all our beautiful wondrous home where we live with great harmony and faith with what we all have to and need to do.

You are wondrous beings, offshoots of our Glorious Father. We would have nothing without his offspring broken from His body to gain knowledge to replenish those souls who have lost their way. Help is needed from all possible corners of all of your souls. You are all able to carry this function to the extreme that is why knowledge to all, knowledge, knowledge and more knowledge is needed for all to succeed in their quest for learning on your beautiful but solid state.

This book is to start you thinking about your spiritual lives. Do not let your lives stagnate. Beauty of colour, of thought, beauty of all ecstatic doings is waiting for your thoughts and input.

Love the writings of this book. The words and happenings are truth, love. Decipher these written prayers as they are sent from God our Father who is pure love and light.

God Be Praised

Words by Archangel Michael
30th January 2011

Hear Our Prayer

Divine God in Heaven hear our prayer; Deliver us from the ways of evil.

Help us to love those around us giving them a prayer of beauty and Divine kindness straight from your heart.

We beseech you to forgive our earthly endeavours that fall to the wayside.

Give us the strength to pick ourselves up and continue our efforts of fortitude for your sight beyond all else.

Help those who are sad and struggling with their lives when a prayer of kindness is for you to give.

Help those dear animals who just want to be loved and treated respectfully.

In all, help the Animal Kingdom to take strength in the good that is offered to them in friendship and to shun the acts of cruelty as they will find love on their return home, those who commit the act of needless cruelty to pay at the end of their life's term.

Dear Father, continue to show us the love we have felt over time. We ask for your blessing encompassing our lives from birth to our life's end.

Thank you for all we receive in your name of Glory. We seek only your love and to find you in your beautiful awesome light. Help us to find you, show us the way to leading a perfect life and to walk the path of Glory to you.

Take away those who cause terror, taking away the lives of innocents, maiming, disfiguring, and causing much grief to loved ones.

Help us all to see the light through your Glorious eyes and allow us to speak with love, singing your praises throughout the land.

Dear God Help Us

Words by Archangel Michael
6Th May 2011

Preface

~✳~

This Book Belongs To Our Lord

The Divine Archangel Michael, Guardian Angel and Guide presented the prayers and words given by our Divine Lord to us all who are living on this earth plain. Thus enabling all to benefit from their beauty, truth and wisdom.

~❋~

Write From Your Heart

Dear Carole, you are blessed with this work given to you by us here in the Heavenly Kingdom. Write from your heart, delving into your spirit awareness, so that other souls can blend into life's spiritual circles.

The earth plain is also blessed. All souls there must be delighted with all they learn and experience. It is only by following this path of gladness that knowledge is gained, helping those souls to find all this beautiful knowledge that was meant to be, to encourage you with your life's plan. You were quite aware of your journey once you arrived on the earth plain. It was your journey laid out for you, and if you followed the correct roads at your crossroads, all would become clear for your journey.

We, above all else, know how difficult a lesson this would be, because when you are offered a different course to run, or an exciting path to follow, you are faced sometimes with a dilemma of which way to go, then the choice is yours. Whether you take one road or another does not mean failure, just that you then must live that part of life you have chosen, before you come across another crossroads for you to make a further decision.

What a wonderful decision you are asked to process. Wherever you go on your journey is another possible success for you to enjoy and live. If the decision you made is not the right one for that time of choosing, it will be wonderful for you to know it, see it, and feel it, correcting anything or all to make your life span clear.

Yesterday was a miraculous time, now is a miraculous time also, but the future all depends on the life you lead then. Wake up your spirituality. You are here because of God our Father; He gave all this beauty and life. He asks that you follow through with interest, relishing all that is given and set before you.

Think deeply as to why you are here, how we work, how our bodies are so wonderful and awesome. There is a reason for every part of the body, but nothing would be useful at all, all this wonderment would be for nothing, if that body had not been given a soul, that soul is you, by our Father.

Think about this carefully. Start your spiritual calling, see with wonderment without complacency. The body given to you will be useless when you have finished with it. We could say what a waste of such beauty, wonderment and a brain more clever than an earthly computer. Our Father in Heaven is happy to waste this body which has protected the soul, because you, as a soul, are loved beyond all else.

Find your path to the light by searching, learning, loving, helping all others who are in need of your help.

God Loves You Always

Words by Archangel Michael
13th March 2011

Acknowledgements

~✳~

It is with gratitude that I mention my teacher
Sharon J Moyle of Cornwall who spent many hours
encouraging my spirituality thus enabling me to
bring to you these beautiful words and prayers

Thank you Sharon for your patience and fun during my lessons

~✳~

It is with so much love that I thank my husband Chris for his patience over many months; Granddaughters, Rosie for her excellent input on the Word Processor and Lucy to include her poetry. Also my family who generously gave their help to me whilst putting together the glorious words given to me by Archangel Michael

~✳~

Requested of Each One of Us in the Angelic Realm, To Give a Contribution

~✳~

Requested Of Each One of Us in the Angelic Realm, To Give a Contribution

Difficulty writing an issue worthy of others to read has to be beyond thinking on a small scale, but allowing a mind to take over a subject that is interesting to read and giving knowledge along the way.

Firstly, there must be a subject on which to base the writings. Secondly, the writer, being able to take on the words that must read with intelligence overflowing as the article unfolds and is read by many a different culture and different gender. It has to flow with so much mystery as to enable the reader to have the need to pick up the manuscript again and again, because as the story unfolds, the interest and excitement of the content must unfurl and capture the readers' imagination.

We in the Angelic Realm are so very interested in helping with your book. It is requested of each one of us, to give a contribution to the manuscript of a placement of extreme interest to you all on earth, us knowing how difficult a life it is to complete there.

We are not able to help you live your life as that is a responsibility purely of your own. We can help you with stories to do with how we live in Heaven to a certain extent, only as much as we are allowed to give.

It is with immense excitement that we enlighten your souls on the way our Father created us all, elevating us to the ways of His love, harmony and kindness, after much knowledge and happiness unfolding into enlightenment, hoping we can pass on to you all for your own goodness, extending the invitation to you all to take upon yourselves the willingness to read our wondrous stories of wisdom, stories of what our dear Lord and Master is like.

To believe in how much He loves you and watches your endeavours with trying to succeed with your experiences and trusting we here with the way forward of your lives on this world of beauty. If only you all allowed yourselves to accomplish world peace, no wars, or terror, but love beyond all else, this is God's purpose for you all.

God Bless You All

Words by Archangel Michael, 2nd April 2011

A PRAYER DEAR
FATHER IN HEAVEN

Dear Father in Heaven, we ask for your help to deal with weaknesses that surround us all. We ask dear Father for your Divine energy to be showered upon us enlightening our lives, letting us see our daily tasks ahead of us, to carry out these tasks with energy, love and kindness, giving us your strength to continue doing our menial tasks, not hesitating at the requests of help showered upon us.

Father, help us to help others along our life's path, generating wisdom and knowledge, strengthening our resolve not to find our load unbearable, just to carry this burden given to us pushing us to our limit, doing this with much love, dignity, and hope.

We are here dear Father to carry your name and wishes forward to all those who have the will to listen, lightening their load enabling the journey in front of them to become lighter as they go forward in life.

Father in Heaven, help the dear little children who are in need of your love and cherish them as they are born into this extreme world. Send to them Jesus to devote His time, His gentleness, and softness and after all else, His love.

Dear Father, take away the wars from our shores, take away the wars from our world. Help there to be peace throughout the land, taking away those who wish us harm. Give peace of mind to those who run others' lives, take away their cruel ways of leading a country. Let them see the way to you dear Father with enlightenment seeing the error of their ways enough for them to love others as they would be loved themselves.

Send to us dear Father your Angels of Mercy bringing with them your Divine instructions for us to follow with all our strength, our fortitude. If we are not sure what to believe, not being as enlightened to your existence as others, please teach us with much patience, sending your Divine instructions encouraging us to follow your beautiful light in love and faith.

Dear Father in Heaven, lighten the existence of the Animal Kingdom giving them your hand in love when they are in need. Enlighten those who have cruel tendencies teaching and showing them the way to their true calling, loving and helping all things.

Dear Father in Heaven, Help Us, Words by Archangel Michael

Angel of Mercy

~✳~

Angel of Mercy

I am here, an Angel of Mercy, so excited at the wording for the book. We here in the world of spirit, the home of Holiness of our extreme God, bless all those partaking in the making of a book solely made up of inspired words written with love and so much strength of mind, just highlighting those thoughts illuminated by an interested party wanting to help all in need on the earth world.

We here in Spirit do not need the help that is asked for by those on earth. Nevertheless you are our brothers and sisters partaking in a difficult test, an enactment on earth and we are so joyous to listen and help all those who are suffering in whatever way we can.

As an Angel of Mercy the plan laid out for me is to offer my help to those who are in need of strength, fortitude, to help the lonely, those burdened with grief, offering my love as an Angel of Mercy giving solace to those who ask in God's name for the burden of suffering to be lightened.

God has given to me enough strength as an Angel of Mercy to carry out His plan for me to work on His behalf, lightening the sometimes unbearable load carried on shoulders not yet wide enough to carry such a burdensome weight. My task is to shower the world with love showing mercy in God's Holy name.

Words by an Angel of Mercy
8th September 2011

HELP US DEAR LORD

Dear Father in Heaven, hear our prayer.

Solve our mysteries of our beautiful world.

Keep us in your sight always.

Help us to search for your blessedness in all things.

Keep us from hurt and cruelty, helping us along the path of your Glory.

Help us to know we are your children and are beloved in your eyes.

Guide us Oh Heavenly Father in all that we see and all that we do giving our all for your salvation.

Love us dear Father when our lives seem to fail, keeping us walking straight along life's pathway to success.

Help the weak and disturbed bringing them in line with your prayers for them.

It is life's plan that we seek your help, give to us your experiences, and give us your love.

You are our mentor dear Lord, send us through our lives with strength, fortitude and if all else fails to take us to our home in your light.

God in Heaven show us how to love from our hearts reflecting your beauty, your love, your Glorious Being.

Help us to want all this with your help above all else.

Amen

14th May 2011

Will You Please Help Me Write A Book?

Carole you are able to write a book with my help that I am happy to give. The Angelic Realm and all who are therein will help dictate when you are ready.

The choice of prayers is decided by our Lord and Father He will take the interest. Think of how you would like to start this going. Prayers will be given explaining our home here in Heaven. Prayers will be given to you from our Lord on High, prayers of the highest order.

The book can originate from you but just by asking the right words will be forthcoming. God in Heaven has high values and will accept nothing less than perfect with his name behind any prayer.

You must start by putting together a beginning of your spiritual journey encompassing all the wondrous altercations you have been given and experienced through time. We here in the Angelic Realm have purposely and deliberately downloaded and sent to you these wondrous acts and experiences for you, to help you through and start your spiritual journey of which you are doing well.

The thoughts I pass on to you as you write are reminding you of notes written as a future reminder of these occurrences and experiences. At each location, each venture, each treasure you unfold will make a delicious insight into our wondrous world passing on your heartfelt experiences to those earthly souls who need extra simple and easy to behold happenings, enthralled into an easy to read easy to comprehend fascinating compilation of beautiful Heavenly and Angelic prayers.

God Be With You

Words by Archangel Michael
27th January 2011

Should I Ask The Question, How Am I Doing?

I felt rather cheeky asking the question, how am I doing. Even so, I did ask, to my surprise and delight received the following message in reply. I then asked the question of myself should these words be included in the book, as they are about me. Well, here are the words that I am so delighted to put on show, because they are truly from Spirit and our dear Archangel Michael. All that he gives to us all, we must accept with thanks, appreciating all we are given in gratitude.

How Am I Doing?

Answer

Here you are Carole, as requested, and yes you are able to ask me this question about yourself. There are some things I would not be able to answer.

Our Father and Creator asked that you were sent to the earth plain to gain experience spanning over centuries experiences regarding feelings of extreme approaches.

You are from level 16 here in spirit which is a high plain to be. You are needed here with these experiences to teach over here. You are limitless in all you do, encroaching your love in life wherever you go.

Join me in searching for lost lives or souls who desperately need their lives fixing.

We, the higher Beings in Heaven, respect and cherish you, encouraging you to continue your life of good. We are proud of you minimising your plight, helping others with your thoughts and prayers.

Continue along your way Carole, helping all in need and when it is time for you to cross the Great Divide we shall meet you with arms outstretched, welcoming you after your life on the earth plain.

We are pleased with all you do. You are respected by us all. You have done well with the plans given for you.

Respect Always

Words by Archangel Michael
17th January 2011

When This Book Is Complete Yes To Book 2?

God in Heaven is delighted that His name is used to put forward words and prayers, allowing more such words to be forthcoming in the future. Carole has already asked that once this book is complete she be allowed to proceed with book 2.

It is with complete favour that we here in Spirit encourage her to do this work.

The wonders of Heaven and Spirit World need to be told to show all on the earth plain, those who are interested, where they will reside on completion of their term of their education. There is little need to cause unhappiness, frightened thoughts, when all will be a journey of enlightened travel as life on earth becomes to an end.

Have faith dear ones; remember from whence you started your journey. You do need to use imagination a little to behold this wonder. You are so blessed in the eyes of your Father. He yearns for the day when all his children love one another, including the Animal Kingdom, what a blessed day that will be.

Each day forth is to bring you closer to our Creator. Think of how it was, how it is, then most importantly how it could be and will be, if all learn about passing to Spirit peacefully, meeting those souls who are looking forward to your return home.

Do not let your mind slip into despondency and weakness. Be strong and know that you are helped more than you know. We sing out loud when a soul on earth is awakened to the light.

Do not doubt dear children, you are loved beyond all else. Speak to me, I will listen.

God Be With You Always

Words by Archangel Michael
3rd March 2011

A HEAVENLY PRAYER

Our Father, our Creator and Mentor grants us favours, stories to enlighten our earthly lives towards our own salvation.

Prayers are said and are a way to enlighten us when things are in a bad condition.

Glorious Light and well-being are sent to all as a blessing, enabling all to see the Light and Blessedness of our beautiful Creator.

Join me in proclaiming our Blessed Lord and Father. Join me in raising future hope to all who follow the light with hope, glory, and faithfulness.

God blesses your journey in all things and sends you access to our wondrous home which is Heaven sent.

Glory Be

Words by Archangel Michael
16th December 2010

Message From Baraccus

~✳~

An Angel Came Forth To Preach

Words by Angel Baraccus

Once upon a time in far off lands an Angel came forth to preach a story to all human beings in order that the words would encourage the true thinking about life existing after death. In order for this to happen Archangel Gabriel sought permission from our God enabling the right wording to follow forth, with the existing wording enlightening all human beings in order that the words would encourage the true thinking all around us all, pressing us with a distinct feeling of beautiful ritual giving the most outgoing lights that can be seen by man.

If it weren't for the Archangel's wording being sent to the front of all those speaking in the different tongues, many saved by salvation would not have been necessary, nor words forthcoming so eagerly waited for in these early years of believing in our Majestic God.

Listen to the words given in peace and loving devotion giving blessings, feeding all with these words of wisdom and natural ways of goodness, leaving the turmoil behind, only seeing peaceful aspects of a lonely domain.

We Angels love to be spoken to with inspiring words, turning simplicity into wisdom, beyond any other that comes before or after the following to the rightful way of the light, lighting our way through darkness into a loving home engaged with all the loving ambitions that do exist even though we seem to let it pass us by.

Love us dear redeemer, dear soul of beauty and discovery of wondrous approaches. Give us the way to this light; let us follow your steps along a rugged road to the way of salvation.

God is love and He asks that you send to Him so much of humility, conscience and Divine knowing, for Him to be so proud of what you have given Him enwrapped in love, serenity, a heartfelt feeling of loyalty, to all others whether they are followers or just waiting for that day when a life will be opened and blessed with the sacrament of our loving and resourceful Father in Heaven.

Reach out to Him, He will answer. You must open your hearts to love, to hear and receive an answer.

Glory Be To God on High

Words by Angel Baraccus. With love to you all struggling to live a useful and Godly life there on the earth plain. Love is with you now and always.

1ˢᵗ August 2011

Message From David, Of The Spirit World

~✳~

Seeking Solitude

By David from the Spirit World

Those who are seeking a world of solitude on the earth plain will find it daunting searching amongst a hierarchy of noise, vice, haste, everything but solitude.

To reach a state of quiet restfulness, you are searching for solitude; it takes a certain peace within yourself, within your mind set. It is to look into your own mind, quietening a healthy body and soul. Meditation is a sure form of relaxation, aligning a body's chakra is another way of lightening a life's turmoil when one or more of these chakras has misaligned due to the pressure of a life.

Allowing life's load to continue pressing with strength onto a human's brain without any relief is a great burden as life continues without unloading the worry and burdensome pressures. Unless these burdens are relieved or released it will be like a volcanic eruption causing the mind of the person to erupt with pressure.

It is for you to find a way of reaching that restful mind, it is another's experiences of meditation written in books purposely to help those who are searching for ways to relax and meditate, giving to them the value of their own experiences.

There are many interesting ways for you to learn how to relax, there are many books on this subject just waiting for you to read, the author's just waiting for you to find and pick up their books they are so proud to have written putting their experiences into words just for you.

A journey into the world of relaxation is such a valuable asset for your mind to own and put into circulation, as you live your life collecting negative vibrations as well as positive along your everyday ventures which are just crammed with worries, each of your days lived are full of these happenings whether good or bad experiences that only you yourselves can say if your minds can cope with the added responsibilities.

Search for ways to relieve the pressures life throws at you along the way, find the information you seek in books of treasured thoughts and experiences, given to us all to share and take into your lives to give relief to the sufferings of soul's who are not able to cope with the normal ways of living on the earth plain. Find your ways to unload the stresses and strains of workloads not realised by hard working people, until their minds are overloaded with information weighing them down, not as realised until the bearing down has become too intense.

Rest and relax find the time for meditation, meditating is so much more beneficial to an overloaded mind even than sleep. Sleep is restful to a body; meditation is important and restful to the mind. Accept all of this as read, it is for us to help you in all ways that are possible, we give to you this information with our love asking that you value all that is given and put to practice. Seek out your information, take in the values that are there for you, put into place the instructions on how to succeed, continue on your way with a thoughtful mind ready to release the strains from your lives.

Having asked who these words were given by, I received this reply.

It is David from the Spirit World very eager to be included.

I am David and I have lived on earth in previous years'. I have been fortunate enough to have my words put into print. There are many here waiting their turn to do the same.

These words by David from the Spirit World
17ᵗʰ September 2011

Message from Angel David

~*~

The Angelic Realm Where I Reside

Words by Angel David

The Angelic Realm where I reside is a place of great brilliance and splendour consisting of extreme light only to be seen by those who abide here. I have taken my thoughts to a higher Being receiving permission to do my best putting into words that which is beyond word description.

The Realm of Angels and Realms of Light are here solely for those Beings whose responsibility it is to take care of our God and Creator. It is not possible for the souls of mortals to travel to these magnificent Realms of Glory where the strength of extreme power and energy radiating the light, empowering the wonders of the universe to exist.

How wondrous is our God and Creator, I ask for no other from Him other than His love giving to us the extreme energy from His magnificence. It is only possible to tell you my experiences, showing to you and explaining to you about the wonders of my home of splendour, complete Holiness emanating from my God and Creator who is to me beyond all aspects of love.

I am permitted to allow you on earth, to all who read this book, a glimpse of my magnificent home still far away from His home, but near enough to be able to explain to you His Glory. However much I give to you explanations in your words, trying to give an idea of the beauty and the energy, the extreme qualities of our high Realms, it is not enough to be even close to the beauty, the extreme feelings of experiencing the great light and pressures to those who are accustomed, to those who reside here, those high Beings who work here with all the love, light and the feelings of awesome joy and perfection.

For you to imagine all that is portrayed by me, to take the words and transfer them in your minds into idea's then imagining in the form of a picture, even then it is impossible as your mind has never experienced anything with such beauty, of such great magnitude as to give to you a clear picture to experience in your imagination where I live.

I hope that when you have read the words from me of this document, you will put your thoughts to further play to reveal all that is meaningful to you searching for the wonders not yet felt by most.

The vastness of our universe with all its wonders, of its beautiful greatness; can be thought too much to involve words explaining its entirety its forever and its eternity.

It is with much love to you all that I share with you my wondrous home in the Angelic Realm just seeking to enlighten you a little further, giving to you the interest we hope you will share with likeminded people on earth. The love that we generate here in our Realm for you all is intensely felt and distributed throughout your world, all that we ask is that you receive it with your love, with your knowing that it comes from our Glorious God and Divine Creator.

Words by David from the Angelic Realm
20ᵗʰ September 2011

After I received these words, I asked "WHO ARE YOU PLEASE"?

"I am David from the Angelic Realm so wanting to help you imagine and understand to a small extent where I live, to give you who have the extreme interest wanting the knowledge leading you to the great unknown, but allowing the knowledge to increase your spiritual awareness allowing you on the earth plain to be nearer to your God serving your own identity as you search, as you walk along your life's path.

I am an Angel who serves my God and Creator first and forever. Seeing the want of soul's on the earth plain it has become evident that spiritual information about the higher Realms nearer to God serves a wonderful purpose enlightening your souls helping you to look for the light."

Messages from Angel Ezmelda

~✳~

A Message from Ezmelda an Angel

I just love the beautiful, beautiful words I received from an Angel called Ezmelda.

Sometimes when I am receiving words I have the feeling that they are not from Archangel Michael because there does not seem to be power behind them. Now and again I will say these words are not from you Archangel Michael, they may not be strong enough words for the book. When I originally asked if I could write a book - can I, the answer was that He and those in the Angelic Realm would help me to write this book, so it is not completely surprising that I receive from another source.

As it happened today, and is written on the following page I received the words that I have named **Parable Another Way Of Saying Story** and while they were just a scribble in an exercise book and did not mean anything to me I mentioned to Archangel Michael that the words were not His and that they were possibly a bit weak (saying it with all the respect I could muster). Then I received the following beautiful words:

"I am an Angel struggling to do my part giving the right words for you to accept with reverence and open arms thinking that what you have been given is worthy of the book.

I ask that you accept these words by me allowing them to be as valuable and as knowledgeable as is wished from me. My name is Ezmelda and I send all love that is possible to you."

Thank you Angel Ezmelda, thank you so very much. I graciously accept the words by you and with much pleasure include them in our book. Without the words from Archangel Michael and Angels like you from the Angelic Realm there would not be a book to share with the world at all.

Parable Another Way of Saying Story

By Angel Ezmelda

It is with ease that I send these words for you to write in the pages of your book. It is a wonderful event waiting for the right words to be given to you.

We would like to send you words of immense beauty and interest to add to those already received over time. In all the books the words written contain words of love, intense love, teaching all those who wish to listen what life is all about.

How many parables do you know of and can remember enough to recall and speak about it openly with all those who have the interest? The word parable is another way of saying story, but a parable is in its own way giving you a teaching of an everyday occurrence, that has an ending that involves helping others who really are not sure whether the situation warrants the helping of others.

So, read a parable and let it help you understand how wonderful it is to just help in a small way, not even involving money in mostly all cases. In the Bible are written many parables that mere mortals will class as boring or maybe simple because they do not have a complicated or difficult meaning to the story, but has so much to offer people telling about their lives and how they can help their neighbours.

God bless you all who seek the truth, peace and the way to God our Father.

Words by **Ezmelda an Angel**

30th April 2011

More Words by Angel Ezmelda

I am an Angel of lower degree putting to you my simple story about water to be included in your book. Please look and approach my words with patience and joy acting on my wish to include me in the words you have been given.

This is Ezmelda once again with much of my love and sincerity for you in your quest for interest and learning.

God Be With You

WATER

It is without doubt the most valuable commodity that God has given the earth, water that feeds the actual living, to drink, to water, to cleanse, all that is encompassed by it.

God made the clouds the winds in order that there is a continuous cycle of water to sustain all that it encompasses all life. Water is rain, snow, ice and this drains into the earth, into caverns and under the earth's crust circling around the world starting the cycle all over again. The sun drying up the rain causing moisture to rise that generates clouds that start the rain all over again.

There are pockets of water under desserts that nomads and the animal kingdom are able to find with their experiences in life. Water is a generous approach to all that needs it for sustenance, without it there would be nothing at all to live on.

As you look from earth to the Heavens searching for worlds similar to yours, those you can see could not naturally sustain life because of the lack of water, so then of course no water no natural area for it to grow.

Where there is no water only dust prevails, just complete dryness. If you were able to travel, much, much further, light years away, worlds able to sustain life could be found, air to breath, plants, trees, a beautiful paradise of land due to the wondrous water that enables life.

Not all planets that are so far away from earth need air to sustain life, but water is a necessity. On some planets life is in the water.

Words by Angel Ezmelda
8th May 2011

Dear Ezmelda it is a joy to hear from you again and with much pleasure I include into the book your simple story about water. Thank you.

Message from Joseph

~✷~

Love Is the World's Great Life Giving Source

Words by Joseph

These special words from me are to be passed on to all beloved men on the earth plain.

It is with great clarity that the story ready to unfold is sent with respectful recourse wishing that all who read it are blessed with the greatest love sent from our Heavenly regions.

Without such feelings of joy and love that we feel to such an extreme, all would be sombre and meaningless upon your earth and here in our Heavenly regions. Love is the world's great life giving source and without it there would be no light just a darkness where we could not sustain life, all would be taken over by miserable depression amid the darkness.

The extreme light here is emanated to you all lighting your very existence, lighting your pathway to knowledge which is the very reason to have earth for your existence. Look to the light and if you are in doubt that it exists beyond your existence, look and learn, take your life's development further than just what you can see along your life's way.

Look beyond, I am there, we are there wishing for a further soul to break through a disbelieving barrier to knowing of a Glorious God and Creator who has created this further home beyond your life on the earth plain for you His children.

From our home we pray for you all, asking that a kind, thoughtful deed be shown to a loved one, a neighbour or friend, a dear soul in need of your help. We are filled with joy just to witness a kindly smile, a helping hand for a soul in need of a friendly gesture, a word of encouragement starting a process of alleviating loneliness in another life's difficult progress.

For you to show such love to another human soul after only a small amount of encouragement, or a touch of persuasion asking that you receive knowledge from another's experiences after pen has been put to paper with words for you to read, think about and process in your mind, enlightening the long journey in front of you.

It is then for you to pass on to others what you have learned, what you have experienced. Not necessarily in book form but by word of mouth to all eagerly awaiting knowledgeable words, with your input of love given to all those who search to expand their minds showing that there is more to life in love and Heavenly peace, this in your words showing the way to the light.

It is so exciting to ask this question - *Who are you please?*

My name is Joseph looking to your world from my world where we reside side by side. I send to you this message with so much love and encouragement hoping that you will read my words again and again remembering me Joseph as a dear friend from home.

Words by Joseph of the Heavenly Realms

24th September 2011

Message From

Josiah

~✳~

Our Dear Gentle Jesus

Words by Josiah

The true place of our Christ's crucifixion is plain to see on the horizon a memory of your imagination. It is so possible to see the cross that caused our beautiful Lord excruciating pain until the ultimate deed was done.

As far as the instigators of the scene were concerned, your dear gentle Jesus the Son of God died on that cruel cross for all to see, some with awesome love in their hearts finding the scene distressing knowing what cruelty was taking place. Most of all because of the innocence of the man they knew as Jesus, the Son of our God in Heaven who had such love and kindness in His heart for all living things.

There were some with no concern as to who it was suffering there on the cross, however innocent the sufferer it is just an outing for them to take in all that was going on around them, not taking into account the freedom of thought and belief and that they may be seeing an error of judgment.

In the Garden of Gethsemane, Jesus after His suffering and death appeared to Mary Magdalene saying to Her that He was now at His home with His Father. All His suffering at the hands of men on earth finished, that it was up to you all there to believe in all He stood for and taught you while He visited this earth for that short time.

Jesus is extreme love and compassion, His words to His Father for those cruel men being the cause of His great suffering was "**forgive them Lord for they know not what they do**". To ask forgiveness for those who had wronged Him in such an extreme way treating Him with disdain and disrespect helping to carry out such a terrible cruel deed to another human being.

It is such an awesome thought that you all have a soul that will go home to Heaven after your lives are done. Jesus told you this before He died, this is how you know that Jesus after His body died, His soul very much alive, journeyed homeward to His Father to carry out an eternity of giving His love to many soul's whose lives are journeying on this planet called earth.

Glory to God on High, glory to Him that sent His only Son to earth for your own good to glorify God's name to teach about His existence and spread His name amongst His people.

Jesus lost His life for you all there on earth knowing that He would return to His Father in Heaven. He tried to share this belief with you all. It is only by trusting in Jesus and His words that it would become possible to see and believe in all He said, that you would believe and know from His words it to be true.

Words by Josiah from the World of Spirit
1st September 2011

I am Josiah from the World of Spirit. These words of Jesus are from my heart. It is my wish that these words with my love are included in the manuscript of words and prayers.

Message From

Little Katy

~*~

A Muddled Story From Little Katy

To be childish and sweet is an expression loved by most and is related to childlike innocence. I am here and I am childish, sweet and yes, innocent, and my name is Katy.

My home in the World of Spirit is where I belong and I am happiest there. My short life on earth was also a happy one until at the age of seven I became ill with a child's disease, leaving my life on earth to pass over to my home here in the Spirit World, receiving a wonderful welcome from all those I knew and that loved me.

"As happened quite often I felt the message was quite weak and I shouldn't continue with it right now maybe I would go back to it later. That is what happened I went back to it one week later and received the rest of the message".

I am Katy (the spelling of Katy was impressed on my mind) I am happy that you have given me a chance to continue my story.

I am here with my mummy and daddy at home. My mummy, daddy and me were in a car that crashed sending us home through time. Mummy and daddy have another boy on the earth plain, a son, a brother to me.

I was said to be beautiful while on earth and mummy and daddy call me beautiful now and say that I am the same as I was before I was brought back to my home here.

It has been a short while since I left my brother, he is fine, I am with him whenever I can be, which is often. Thank you for writing down my story.

With love, Katy x

"I doubted everything at that moment, myself, and Katy, how could she have passed to Spirit because of a childhood illness and in a car crash as well. I needed to look into this further and asked that question, what is going on Katy? I need the real facts if we are to continue".

I am here, it is Katy. It has been very difficult for me to remember it all, I am at the moment nursed because of the way my life ended on earth, it was so sudden and not long ago.

My mummy says I was very ill and was in the car on my way to hospital with mummy and daddy and that is when the accident happened.

"Thank you Katy for giving to us your story, we hope that you are completely well soon".

Words by Katy living in the World of Spirit

15th September 2011

Message From Percious,
Of The Spirit World

~✳~

To Imagine Us Without a God

Words by Percious from the Spirit World

The love of God is beyond any other conversation, it is strength and beauty and can be talked about with any person in any environment whatever the circumstance. It is a subject of deep importance, enlightening any progress made by discussing such a beautiful subject.

God is love; each and every one of you must come to know this however long it takes. To imagine earth without a God to believe in, to discuss and know that because of Him we can rely on life after death. The word death would be used by those inexperienced of our God, that to die is all there is, as we try and imagine dying, that it is forever and eternity, just nothing, try and imagine there being just nothing.

We are too beautifully and cleverly put together, so well thought out by a Being, our God, out thinking all that there is, out thinking the possibility that someday a computer will try and out think humanity. That is why our God is who He is, our Maker, creating all that is in this world, the universe and our Spirit home, Heaven.

We can only be thankful that our Creator is love beyond all else, that for those of us not perfect have more than one try, or more than one chance to get it right. So for Him, we can learn and experience faith, be thankful for His utter patience benefiting us while we live here on earth asking Him for help when in despair. So, if you do not believe in God, but when in utter despair pray to Him and ask His help what does that mean? That you desperately need for His existence to be true, that there is an afterlife there for us all, that home, our previous existence is there for us to pass over to, from a finished life on earth.

It is the difference between complete end, death, complete nothingness, or love, beauty, a Heavenly home, where our loved ones whom we love beyond all else, our children, our family, that when for whatever reason their lives end, we know in grief that we can be joyous as they are at home, in Heaven with other loved ones.

To be then perfectly honest with yourselves, that our life, our journey to earth was purely an enactment, a visit to the earth plain to gain knowledge to take back with us, to give experience not only to ourselves, but to all those at home in Spirit eagerly waiting for heartfelt feelings, experiences all from a difficult place of learning.

God Bless You All
26th June 2011

The above message was sent to us from the World of Spirit

It is Percious, I am from the Spirit World, it being my wait to be involved with this precious book. These words enabled by me to try in my small way to add my account of what I see and hear to put into words interesting enough and learned enough to be of help to all those who read this book, these pearls of wisdom.

Message From Peter, Of
The Spirit World

~✳~

The Middle of the Story Is Up To Us

By Peter from the World of Spirit

Dear generous ones wishing to learn all you can about the Spirit World and beyond. We are here to help you find out about God's ways feeding His children and coaxing them in the right direction to deliberate over their minds turmoil, when what learning they find out along the way becomes more than they can take in at first point of call.

More often than not it is not believable at first try and then comes the wonderment when it all begins to unravel, as does a story, having a beginning and an end, what happens in the middle is the main feature of any great story or tale.

We here in Spirit watch in order to keep up with the life stories enacted in episodes. It is an unbelievable adventure that has been purposely given to you all on earth for you to learn and for us to learn with you, even though it is more than possible that you slip or fall along the way.

Finding a partner with whom to fall in love, now that word love is to encompass a beautiful deep feeling, but love can turn so quickly into pain and unhappiness, we experience this too. To find a partner, fall in love, marry and bring children into the world starting a further cycle of life, bringing soul's from our God into earth's circle of intense happenings.

To deliver words of Divine wisdom along with these thoughts meticulously thought to match the deed is a difficult enough scenario. You watch a play or film, we can watch with great interest a life's story unfold and not knowing the ending until it is acted out properly by those involved.

It is not an easy experience for those souls who are the actual participants in the scene, pain, love, lust, jealousy, hatred, cruelty, what a mixture of experiences to watch and try to examine the pattern arising from all that is occurring.

Although we can watch as explained, it is not possible for us to say whether the occurring story is acted out correctly or not, it is not for us to comment. Living a life is difficult a mission to be enacted as is possible to have; some soul's doing better than another in one thing but quite the opposite in another. It is possible though to think what I would have done to handle that situation.

Sometimes we feel we have the solution and if that were correct, how easy it might be just to nudge the main participant and just say, I have the immediate answer for your problem, I can solve the mystery from afar. Of course this is not going to happen as the story is yours and what a wonderful job well done for the most, but for many others it is such a struggle to solve the problems of life.

Upon your return home you will go over your life's history enabling you to say that was successful, or perhaps, well, I could have done better with that. You are not judged to the extreme, just your life's story discussed and a remedy found because of that discussion. Although these happenings are serious in many ways, it is possible to not only applaud but to laugh at an event that looks and feels funny and laughable. We love that because laughing is so important for lifting the vibrations and energies of us all in Spirit as well as the instigator.

Knowing that you are to be asked to look through life's history can make it a very nervous participation, but you are asked not to feel that way because it is known that an experience on the earth plain is more than just difficult. Even though sometimes you fall by the wayside you pick yourself up and continue spanning your life's journey with enthusiasm where it is possible at that stated time.

We here in Spirit laugh with you, cry with you, we applaud wherever possible and we live in your hearts, encouraging and send to you our feelings of love although you will not always be expecting that feeling, whether that is because you do not believe at that time or that your thoughts and feelings are elsewhere.

God be with you always in your hearts. You are loved beyond all else.

Words by Peter of the Spirit World
12th May 2011

Once again with excitement I asked who it was giving this wonderful message, this was the reply:

"Indeed it is your choice whether to ask about the sender of this writing and I am filled with joy with my reply.

My name is Peter from the Spirit World hoping to be included in your book. It was necessary for me to ask for permission and joyous am I to have been affirmed to help my Master's information to be passed on to so many others' in need of this loving confirmation on earth.

Thank you for accepting my words for input in the Glorious book of God."

I am delighted to include your writing into the book Peter, thank you.

Carole M. Day

I Asked Peter to Tell Us About His Life

I asked Peter if it were possible for him to tell us something of his history and these are his words:

"My name is Peter; I am here in the Spirit World and have been for many of your earth years. Yes, I have lived many lives on earth doing various jobs living various lives, some quite successful, some not so well. Even so, I am proud of my proceeds of lives, also I have gained much knowledge in my journeying that I can bring forward when necessary, enabling me to relate to people like yourself and others living your lives now.

I am a teacher here at my home, teaching others of my experiences on earth when I existed there over many lives. It is for me very important to watch earthly lives in process as the time goes forward, and as previously mentioned an enactment of ways to live a life. Then I can relate to the play however difficult or easy the beginning, middle or ending is and can tell my students of the occurrences, although the missing part in their learning's is they cannot feel the extreme feelings with which to include in the scenario of life. Knowing how experienced a teacher I am it is possible to give them insight before the play for them is enacted in reality.

If I am requested and by necessity to revisit earth to help in its troubles and turmoil's and knowing that an input by me would be totally helpful, then I would need to think upon the difficult task in front of me and accept what has been asked of me. I do think that would be far off as doing what I do now is relevant in all things."

Words by Peter from the Spirit World
13th May 2011

52

Message From Roger,
Of The Spirit World

~✳~

On Earth to Help the Sick

Words by Roger in Spirit

Medical philosophy is a topic that people do not think about only if it affects them at any one time. Our existence with an enduring life can be while healthy, a life that continues on without illness of any description, this being very fortunate for some.

The Spirit World holds all kinds of doctors giving their experiences that were once used on the earth plain to those who practice medicine now on earth, doing there all to obliterate all the diseases known to mankind, working in unison with Spirit engaging those on earth.

Those souls whose destinies were meant to be surgeons, doctors, nurses, all caring for the sick, were brought about by their love of humanity, bringing a longer life to a suffering body, a suffering soul; it was their choice in life.

We here in Spirit are desperately watching the sick and the suffering, also the carers struggling to minimise a patient's plight. For us it is between causes that gladden our hearts, those dear souls who's bodies are either cured or at the very least their pain managed enabling life to continue on, giving life the freedom hoped and longed for to continue on their spiritual path fulfilling their destiny, their reason for visiting the earth plain.

The other extreme, is the nonlife giving side of things when the soul passes over to Spirit, because the illness of their body made it happen.

It is the surgeons, doctors, medical practitioners, or those who are nursing; it is their reason for being on earth to help the sick, to find out the reasons why these diseases of the body shorten the lives of souls. In most cases it is at the request of these soul's experienced in medicine who reside in the Spirit World, to return to earth with whatever they have learned to help life on earth, to help those suffering, to put into action any additional brilliance that might be used to aid suffering on our beautiful but intense planet.

The following wording was given to me by Roger in the Spirit World.

My name is Roger from the Spirit World. I was a nursing assistant in my last life and my experiences learned whilst there has put me in good stead to assist with medical problems there on the earth plain.

I lived my life on earth for seventy two years born 1901 passing back to my home in Spirit by natural causes.

To nurse those suffering with an untreatable illness was sometimes hard for me to bear with the necessary seeing of life ebbing away from a patient, but also the sorrow of loved ones standing by. The relief and happiness I felt when a patient I was nursing recovered with my help.

It is with my complete love that I give to you these words of testament to add to the Book, giving I hope beautiful wording to help you all on earth.

Love and Peace Be With You
4th July 2011

Message from Angel Samantha

~✳~

A Chorus of Angels

A chorus of Angels in the Angelic Realms is so wondrous an occasion because of the vibrant and beautiful Angelic voices that sing the praises of our God. The Heavenly music and choirs are beyond the knowing of souls on earth; sounds of the Angelic voices singing are of the highest vibrations.

Sing of our dear Lord, sing His praises to Him, tell Him of your love and tell Him of your learning's on how to dispel evil from your world. Sing of love and glorious happiness, you will see the light shining through the mist of unhappiness giving all that taste of our Angelic feeling in all things.

We are love above all else and ask that this love be shared with all those souls who wish the same, only to enjoy life's beauty, maturity, helping with wisdom to be spread amongst you all.

God is love, He loves His children giving them so much love from Him, keep this above all else through your lives.

Words by Samantha an Angel
9th May 2011

Message from Saviour,
Of the Spirit World

~✳~

Living Of All Life by Saviour

It is mine charge here in Spirit to look upon your lives as a bountiful exercise in the living of all life, trying in all aspects to do a wondrous learning of our experiences when putting pen to paper in order that all that is written can be taken to our Blessed Lord for Him to see our way forward doing the work given to us, by Him, on this earth.

It is an exercise for you to accomplish on whatever level you have managed to do. We here in Spirit watch with excitement, with great interest, wondering how we watching could accomplish the same level of strength coping with such pain with life's problems, hoping that the feeling of satisfaction and love will override the whole performance enacted, likened to a stage, we being the audience applauding when the whole scenario is enacted with a pleasing settlement of the earthly play.

We know beyond doubt that you have a very difficult arena in which to enact your life's journey starting from birth, difficult for some, wishing that a strength beyond that of weakness can deliver a dear child's life, giving that child a good start beyond all else.

The difficult overriding part of a soul's destiny is not to know where they originated. If they were aware of such a wonderful home existing they would be only too eager to return, therefore, the whole scenario would be a waste of time on the earth plain. To not know of a home before life on earth sets the feeling in motion of believing in God, in His son Jesus, just thinking and believing however long a time it takes, however many times that belief fades, the experience of the beautiful thoughts that all might really exist is a way forward to a life of complete worthiness, of helping others with stories of your happiness with belief.

It is an awesome opportunity of help for us here just taking in all the experiences turning into knowledge of great worth, enriching our spiritual gain in the Realms of our Father.

I had the feeling that these words were not from Archangel Michael and so asked the question "Who are you please giving these beautiful words."

Who are you please?

Carole M. Day

No it is Saviour, I am in the World of Spirit, enriching both our lives with information, giving wonderful hope to all that read my words of increased effectiveness, giving you all strength to continue your lives with fortitude, knowing you are not alone as your lives unfold.

Glory Be To God, You Are Loved

Words by Saviour from the World of Spirit

22nd June 2011

Message from Sebastian, Of the Spirit World

~✻~

What If Love Did Not Happen As God Envisaged?

Words by Sebastian from the Spirit World

What if love did not happen as God envisaged? What if all was cruelty? This way of extreme thinking is asked of you, enabling you to experience horror, darkness and all feelings lacking the light of our beautiful home here in Spirit.

Love is everything, it is impossible to imagine life on earth or the Spirit World just in solitude and black density. We here in Heaven ask that you go forth with your existence, looking and seeing the light that gives a life worth living with all the beauties that are there for you to ponder.

Instead of looking down at bleakness, do the opposite and look up to the light never having to shade your eyes to this bright light. It is not the sunlight, but a Godly, Heavenly, effervescent light that holds the love and blissfulness that is promised to you from the loving, beautiful, extremely loving mind and home of your Father.

This is a lesson for you, simply asking that you seek the light ignoring the darkness, the pit that you could fall into.

Please care for my words; it is for your happiness that I send my love to you with these words.

God Bless You

I am Sebastian from the Spirit World intensely looking forward to having my loving and truthful words listed in the book. These words are written in truth, above all else read and consider the words written therein.

22nd July 2011

Message from Angel Zeb

~✳~

How Else But By Words Can His Wishes Be Passed On? Words by Angel Zeb

It is with much love that I write to be included in the writings. God in Heaven and His Son Jesus Christ are watching and reading with interest the wording on the pages written so far.

It is with excitement that we know our dear Father's wishes will be sent throughout the world to many people encouraging them to read and then know what is wished for you from Him.

How else but by words written on pages can His words, His wishes be passed on to all. They then have the chance to act, believe and carry His wishes to all else, spreading the wondrous words which are truth and intensity. It needs for you to read and to love the Holiness of His words, spreading the knowledge that has been given to you in His name.

This book is a gift for you sent by your Father. He has permitted all involved to add their words and prayers, beyond all else to go forth into the great world to encourage those who are erring to think again and return to the correct path, where there is beauty and fortitude above all else that you are aware of.

It is the erring of some that sends your world into turmoil. The more who read these words will set the wheels in motion for us to seek an easier way to live on this beautiful earth. Read, take in and pass on the knowledge that you have just gained; knowing that your interest is noted by those watching with true love and trust, that in the future there will be a mighty change to your world.

Love Is Beyond Doubt In All Things

8th May 2011

At the start of the previous page the words started with *"It is with much love that I write to be included in the writings."* I thought yes, that is just wonderful but who are you please? And the reply was:

"My name is Zeb; I am from the Angelic Realm, this of course meaning that I am an Angel. I ask that my words be included. It is with wonderment, with love, that entering my words would be Glorious in my eyes to further this cause for our Divine Father, it would be my entry of light to serve my God."

Words by Angel Zeb
8th May 2011

A BEAUTIFUL PRAYER

Dear Father in Heaven.

Hear our prayer.

Take us into thy Divine arms; lead us through your pathway of healing light.

Bless all those who seek your light.

Bless those who do not, but help them to find the one and only path to thy Divine Home.

Lord teach us all that is acceptable to you.

Show us the way forward generating Glorious love in all that we do in our lives.

We thank you dear Lord for the beauty of our world given to us by you.

Teach us to appreciate these wonders, all we are given in love and joy.

Show us the way forward encouraging us to show love and kindness to our brethren who are so in need of help.

Love us beyond the love we know and give to others. We have such a great deal to learn by using this Divine feeling; we will then know more of what is needed from us in our earthly lives.

Show us the way dear Father, let there be no barriers between Heaven and earth.

Do not give up on our earthly plight but show us the way Oh Lord.

Words by Archangel Michael
April 2011

Archangel Michael,

Please Answer These Questions

~✳~

What Is It Like In The Angelic Realm?

It is often asked what it is like in the Angelic Realm. Well it is a place of deep love, endearment and of Glorious light and energy. It is a place where our dear Lord resides, where He is Divinely worshiped by those Angelic Beings who attend Him and keep Him safe.

The Angel Realms are a serious and wonderful place. They are home for God our Lord and Creator's Holy Beings who tend, guard and house these wondrous guards. Beings who have been chosen before all else to take care of all Godly and Heavenly venue's.

The Angelic Realms house Seraphim, Cherubim, Archangels and all Angels. It is not possible for soul's on lower levels to visit this Realm because the power, strength and powerful learning's are here for God's journeying only.

The energy and beauty is second to none. The colours have not been seen other than Angelic Beings. Beautiful Angelic voices caress our dear Lord. We show our love to Him, guarding Him with all our being, surrounding our beautiful and Divine Lord and Master and Creator of all, wrapping him in faith, love, song, and colour.

He is wonderment. We have been created by Him, creating beauty, management, all things revering Him our Lord.

Ask for a message for our Lord and we here wondrous Beings will illustrate that message with beauty, love.

Gather your thoughts together think of many Realms, think of the highest Realms there are to be found. It is not possible for souls to visit here but we on request, can and will visit you.

Angelic Realms are pure light, power and love beyond reflection.

So much for our God in Heaven is here. Light, light and more light is here. Holiness beyond your thoughts is here.

We Are God's Love
Words by Archangel Michael, 14th January 2011

Do Archangels And Angels Have Wings?

What a question to ask Carole, that answer would give away an Angelic secret. Our Divine Father in Heaven is willing for you to be given the answer to that question because it seems to be a much talked about subject.

An Archangel has wings, wings that are so very powerful, not only in strength but the look of Divine energy and enormous ability of movement, giving the generous behaviour of one so Glorious, a look of Divine Holiness, peace, gentleness over the extreme strength, over all else.

Angels do not have the enormity of strength as do Archangels but none the less they are powerful, loving, generous Beings who work for our Creator giving their gentleness, fortitude and loving commitment to all souls in need of love and gratitude from another area of life.

As with everything Angelic, the choice is given from God on High that they may expel their wings if they feel the situation might for example be visiting a soul on earth in a general sense, maybe in a street or place where magnificent powerful wings would cause alarm.

Archangels and Angels are as you would expect to see them. They are there for you under the instruction of our Higher Lord. They are goodness itself, beauty beyond bounds, Holiness, all things of strength and beautiful rays illuminating all who come into contact with such a Divine Being.

Blessings Be Upon You Dear Children of God

Words by Archangel Michael
28th April 2011

How Am I Doing With My Life?

Answered By Archangel Michael

It is a human failing to sometimes doubt yourself, this doubt as to whether I was fulfilling my time on earth as planned happened to me. A program on the TV was handing out prizes to those wonderful, thoughtful people, who did kindnesses for other people without thinking of themselves. So, I asked my Guardian Angel if I was doing anything of consequence with my life. I am so delighted with the answer to my question.

To receive confirmation and proof that our God and life after death truly exist is the most beautiful experience, the most wonderful of words that could be received. The wording following this short address are to me extreme proof beyond doubt of God and Heaven existing.

It is for you to read these words sent to me by Archangel Michael my Guardian Angel. I truly hope that you can read them and find the belief and proof for yourselves. These words and prayers were sent through me, but so clearly meant for you.

Answer: How Am I Doing With My Life?

Carole, you are of strong complexity, being encouraged by me your Guardian Angel, I and I only will tell you if you are lacking in your interests with your life on earth.

I am extremely keen for you to continue as you are, giving the people on the earth plain words from our God, our Father here in Heaven. Without the input of a strong character, a strong soul on earth, able to pass on by word of mouth to those people on earth who are struggling in their belief of what the world and beyond holds for them, it is you who can put our words of love and strength, the story of our Divine Lord and how you are expected to live. How the knowing of the wonderment, you are able to know there is still life when your life is done on earth.

You have been put on this earth for a specific reason of learning about where you originated here in Spirit. By writing the wise and beautiful words of love and wisdom sent to them by a wise and loving Being here in Heaven will these souls throughout the world read these words and prayers, allowing them to think and inwardly digest the very thoughts sent to them by you, through me.

What a beautiful, wonderful occurrence to spread throughout our dear soul's in need of assurance, peace of mind, Divine assurance that those dear loved ones, all those soul's passing on through reasons not of their control, are welcomed home to be served with the love and understanding that is completely expected here in Heaven.

Do not Carole; wonder if you are doing what you have been chosen to do, because your love for others, the capacity to love and communicate with us here is as asked of you. Your purpose being here is not complete; you have many more words to add to your manuscript for the world to see, words passed to you from the Highest of High. Welcome this with all your heart and being, we here are at length, waiting for a turn to help the people of your world with our Divine words and thoughts.

Words are mightier than the pen; it is by the pen that we pass to all our messages. We send to you our love and encouragement for the incoming words and voices.

Words by Archangel Michael
17th July 2011

Heavenly Laughter - Do Angels Laugh In The Angelic Realms?

We laugh a lot here on the earth plain whether we tell a joke or see something that makes us giggle. Laughing really does lift spirits; in fact, it makes us feel better if we are down in the dumps.

It occurred to me to ask Archangel Michael if they laugh in the Angelic Realms, or in the Realms of Spirit. Do they think what we do or say, here on earth is funny? Do they think there are things to laugh about in their home? Do they have a joke? Do they laugh, if so, at what?

"Yes Carole, of course we laugh and find things funny, especially here in the Angelic Realms as we watch over all you do, if you laugh, so do we. It lifts our spirits beyond all else to see happiness on earth. We do not actually tell jokes here but listen and try to understand yours. We can find that quite difficult, but just seeing your faces, sometimes hearing the uncontrollable laughter raising the vibrations is just wonderful and uplifting for us to pass on to our Father to understand and be joyous.

As we have mentioned before, Divine Beings in the Angelic Realms have not lived on the earth plain, we have gained knowledge just by being with you all as a Guardian Angel or Guide. To understand a joke goes quite beyond our understanding for us to see you laughing at something we cannot relate to.

The Angelic Realms are for us a wondrous place to be. Joy is found for us just by working for our dear God in Heaven. We work continuously helping to encourage your spiritual awareness for Him, whether on earth or other planets.

You ask what we do here in our Divine home. Beautiful choirs of Angelic voices sing our Father's praises. The feelings we have are joy and love above all else. The love we feel for the soul's on earth, singing joyful songs for those who have become enlightened spiritually, opening their mind to God our Father in Heaven. Divine music and Angelic voices go beyond your imagination.

Those souls at home in the Spirit World have all the experiences they need to laugh out loud with your antics, jokes, stories and can relate to things they see that cause such humour and laughter."

I would very much like to infill with more information but I think that is unlikely. I have been given an answer to my question, that is all I am told, and what I am given I am truly thankful for.

How Was The Universe Created?

God created the universe as God created all about you, giving Him a wondrous place to live and educate His children enabling His All to multiply.

Before He created this wonder of learning, our Lord God was indeed the Universe, but He needed to create more for Himself expanding His Divine following, giving them a wondrous job to do, looking after not only God now but His children, making an expansion of Himself ensuring that His children were of His calling, without avarice, without hatred, without jealousy, giving them perfection, throwing aside all that is rife on the world of earth.

The universe is forever, adinfinitum, it is not ever possible to travel to eternity that is for the universe itself only which is our God. Just to imagine that which is never ending, that which goes on never coming to an end. That distance is not comprehensible; you will cause your brain to overload just trying to gain the answer which is impossible.

Somewhere out there where the universe never stops, there must be more like your world with living things, possibly like you to look at, but probably not. Just think of all those exiting possibilities.

You have already been told that there is not just you in this universe, that above all else must make so much sense. Just keep your minds open to such thoughts and teachings allowing all that you have learned to come forth, setting your mind thinking of all there could be, all that there is.

Words by Archangel Michael
28th April 2011

Please Tell Me About The Realms Of Light?

The Realms of Light that is your home demand the utmost respect and high responsibility. The Realms of Light are a wondrous beautiful place and as the name suggests are overcome with such brightness with extreme light that only those trained over time and have gained such knowledge are capable of succumbing to the great light that surrounds us all.

Your home, encompassing this great Realm of Light is waiting for your return enabling you to relate all that you have learned while living your life on earth. The Realms of Light hold such beauty, all living Beings shining and sparkling with the light created by our Father. Earth people could not live here because of the very high energy level that abides here.

This high energy that we live and experience, is not on the same strength of energy and beautiful light that as you become nearer to our Lord, is so beautiful and the light so dense that the wonderment of the power is beyond most Beings. There are Beings that serve only Him that can be in His near proximity.

You are a teacher here Carole in the Realm of Light, you are respected by us and we cherish you, awaiting with open arms to welcome you back home to hear of your travels and experiences. We know that you can and will relate the mixed feelings that all there experience, we understand they are extreme in feeling.

Everything we have here for example, trees, plants, flowers containing colour beyond your comprehension, it is impossible for you to imagine but of course you will remember in the future upon your return. The Realms of Light are vast, we may travel where we want to just by thinking the thought.

There are schools or academies, they are for such as you to relay your experiences in real feelings, to those who know not that which only you have learned and felt the extremity of other things other than love. Therefore, it is difficult; we cannot feel hatred, grief, and jealousy as we are love in all things Sacred and Holy. We do need to know as much as we can what humans, animals, trees and plant life are experiencing for us to give our help where we can, when our Father requests that we do so.

You are asking us for help with publishing this book, what you should do? Where to go? You are mostly asking will the book be successful, adding that with God our Father and Archangel Michael behind the prayers and writings it can be nothing else but successful, being shown throughout the world.

It is with God's instruction that the prayers and wording be given to help enlighten those on earth that wish to learn and become beautifully enlightened in the truth of His stories, prayers and parables all written for you all to see.

When the book is complete it will be published with ease getting the help from our wondrous Archangel Michael. The world is very much in need of words to comfort, to give instruction but most of all to enlighten you all to what happens at your life's end.

Read this book and its teachings, you must feel at peace with faith that life is never ending; here on earth is not your home but just a stepping stone to your real home and life.

So, Carole, the Realms of Light and the Beings who belong there have many jobs to do in the name of our Father. It is one of these endeavours that you are doing now with all the strength you can muster. We wonder over you, watch over you in your ups and downs and help with your endeavours when and where we can.

The Realms of Light are surrounded by the Realms of Glory a step nearer to God our Father, it is a wonder to behold, love beyond all you would know. You have forgotten as is made to be dear Veran, continue with all you do in God's name, helping to spread all He is and what He is, our Lord God our Creator.

You Are Beloved

Words by Archangel Michael
4th May 2011

What Is God Like?

A loving God, a purposeful God. He is genius amongst men, such a powerful force, a gigantic energy that has to be to generate such force, such wonder to create and have created the universe, worlds, everything, however big or small, all was generated from Him. All that is, we marvel at, knowing He is our Protector, our Being our Father.

Without God our Creator there would be no light, no lives, nothing that is even symbolic of what might have been. We are for Him what is, what will be always and forever. He is not as man, more brilliant in all ways not to be imagined. He is pure energy, pure light that cannot be imagined.

God our Father loves us beyond all else as a Father as we are part of His wonderful solitary Divine energy that will stop at nothing to keep us following His pattern.

Our Creator would be seen as pure sparkling light, His brilliance cannot be imagined. His knowledge of all there is can be said as brilliance beyond all else. Never will there be a person nor a computer built by such a person who would be able to surpass our God in Heaven who is Master of all.

Accept His love given to you; never forget from whence this came. Look to the light always, feel His love, feel His energy. He does not seek gratitude from you, just love for Him and His Beings giving solace to those in need more than yourself.

It is impossible to imagine our dear Father and what He looks like. We cannot explain this to you, all is beyond your reasoning, beyond trying to explain. It is impossible for you to comprehend such brilliance such beauty, such knowledge.

Rest assured He is wonder beyond our reach except for imagining and knowing beyond all else that He exists for us all, sending His only Son to meet with us as a proof of existence.

It is not possible for Him to show His wonder to us here on the earth plain, our world as it stands could not withstand the enormity of His energy.

Then it would not require you to have such belief as is needed when all you have is ancient books, modern stories to read and give your mind and belief in such a wondrous real occurrence of time and generations of storytelling.

Glory Be To God

15th April 2011

Where Do Animals Go In Spirit?

It is beyond us all to consider where our dear animal friends reside in Spirit World. This all depends on whether this animal friend was loved beyond all else and would be made welcome in a home of its choosing.

There are many descriptions of non-human soul's, some belong to us in our home, some in bushes, trees, underground, or in water, that are not loved or thought of in any loving way, therefore you ask where will their home be? Also we must mention insects, trees, plants, flowers, all these are purposely made to make the earth world as beautiful as it is and they all have a home in Spirit World.

All living things whatever they are live and breathe, multiplying in their own way, passing to Spirit when their time is right and their lives over. Here in the Spirit World is a supernatural place where all non-human beings have a communal home where they all live in their own environment together, living lives as they would feel is right for them, knowing they are near to their Creator as we all do who have lived a life.

To show your love to any animal, to treat them with kindness enables this animal soul to choose to be with us, or to join their likeness in a communal community.

All God has made He loves, welcoming all His prodigies back to His fold.

Words by Archangel Michael
18th April 2011

Angels Have Their Say

~✳~

Angels Are Extreme Beings

Angels are extreme Beings working above all else solely for God. Nothing else is as important to them as our Lord God.

In His planning of all that is beautiful and awesome, all that is needed for His own service to accomplish His thoughts of life, existing from the whole existence of creating world's containing life originating from Him. He created all that is His in His own light. He created it in complete love setting a world of existence beyond the universe, beyond Heaven.

God created a hierarchy of wondrous protectors to serve Him well. Many of us are aware of Archangel's and Angels, but there are many of those awesome helpers who are nearer to God than any we are aware of, doing His bidding, protecting our Lord, being their sole reason to exist, serving God.

In the beginning, Angels were created by God, and given their own will, their own minds and freedom of choice, those loving our Lord giving 100 per cent of their love for God. Their reason of existence to love and portray all that He is, protecting Him with all their strength.

Where there is good there is also bad and the Angels created had choice, some choosing not to serve God but wanting to lead their existence working against Him. These rogue angels were cast out of His existence and His beautiful world of home.

The hierarchy of Angels live in the Angelic Realms respected above all else for the awesome work of love towards our Lord. There are no souls who live in the beautiful Realms of Angels, but to imagine the peace, love, grace and Divine feelings of happiness, that only the Angels serving in their own Realms know of the wondrous beauty that exists there, we can only think of Heavenly venues and sights that we can only dream about.

Words from Archangel Michael
23rd June 2011

Visiting Angels

I am sure that all visiting Angels to the earth plain are so proud of how they accomplish the Holy lessons and accomplish all to help us to benefit from their strength and their beauty, with kindness a supreme feeling, helping the world to succeed in the quest of greatness nearing the love of our Father.

An Angel is an unseen Glorious benefactor of a highest Realm such a wonder for us to behold. Many experiences have been known where miraculous occurrences have taken place, when we on earth have clearly needed the help of one of these Glorious Higher Beings.

The magnificence and glorification of such a gentle Holy Angel who has knowingly been sent by their sole Creator and wondrous God to work their miracles for life's progress where gentleness, special love, trust and encouragement to believing all that is true, sent straight from the very heart of such wondrous Holy Beings serving one God.

It is generating the very work needed to help with the task of an Angel whose sole responsibility it is to sow love amongst men, giving to them a purposeful meaning to their lives, giving hope amidst grief, unhappiness and life's turmoil's, sourced from that Divine love we know as God our Father.

Bathe in the Exquisite Light of Your Glorious Angel

7th December 2011

We Angels Glorify Him

Feeling the joys of living in our delightful Realms of light of my brothers and sisters, who are Angels of God, living a full and ecstatic life taken up solely to generate glorious love and prayers to sing to our dear Lord.

We glorify Him and sing praises, rejoicing at His existence of wonder, beauty and illustrious world of Heavenly beauty. We Angels have a sole reason to exist that being to serve our God and give glorification to His existence, awesome strength and power generating all that we are in the Angelic Realms, also generating the awesome power to feed your small planet, giving your lives here an energy, giving to you life.

The Realms housing the great Beings of Archangels and Angels are surrounded by beauty that only those residing there can experience. It is not possible for others of a different gender to be included. We guard and serve our wondrous Father, choirs singing music of serenity which is beauty to those who hear.

We watch with great enthusiasm and extreme awareness the activity of lives being lived on the earth plain. We are surrounded by love, we live and generate love.

The Wise Ones who are Beings of great and extreme wisdom and knowledge watch the lives being lived, knowing whether the life has been well lived or if another try might be thought of in some future time.

The energy that is received here in God's Realms by those souls praying for God's mercies and His help for themselves or others, is a beautiful sight, bright lights and sparkling Divine energy highlighting the wonders of faith shown from the prayers and hymns from those believers on the earth.

God is all powerful and extreme love, His authority immense amongst all He has permitted to exist with His words on their lips, spreading that word to those non-believers who must benefit by seeing His all-knowing light, helping all to know our Father and God of us all without doubt, truly exists.

Words by Archangel Michael
28th August 2011

What Is an Angel?

An Angel is a Divine Being whose purpose is to protect our Father, also to carry out His wishes from the Divine source of love and beneficial observations given for us all to take notice of and follow without hesitation.

An Angel is a beautiful energy of extreme strength of complete love. His energy once given to you to be used in their light of unimaginable brilliance, lending to you their soul, only permitted by God if the receiver is respectful, mindful of the greatness of these wonderful great Beings. It is not for you to know the treasure you sometimes hold.

A change in your leverage the change of a troubled perspective can enlighten your peace of mind, change your way of living when so urgently asked for by a troubled soul.

God is love, for Him to send to you such a wondrous servant so close to His world, so treasured by and the intense love they share; defend with all your existence the mighty Angels and Archangels whose occasion it is to love you always.

An Angel's love is beyond doubt the strongest of Divine feelings, giving off that light as that of a beacon searching for a reason to save soul's with that brilliant light alight with flames lit by God, for an Angel to carry forth that banner of extreme light, giving off more of God's love to a fading purpose just for a moment lighting the way, helping those soul's on earth to follow their path just using His light to see a clearly marked pathway.

An Angel is a powerful Being, carrying out the words of God, instructing all of their might to carry out Angelic laws of their own world, bringing awesome wonder to lowly souls about this earth plain.

You are loved, truly loved

These beautiful words by Archangel Michael
4th November 2011

After receiving these wonderful words answering the question "**What Is An Angel**" I asked about the sender and received the following answer.

"It is your Guardian Angel Carole, these words are sent with grace and favour for all of you on earth wanting earnestly to know of we Angelic Beings serving our Master."

Love Is Beyond All Else
5th November 2011

Words of Meaning for Us All

~*~

A PRAYER FROM JESUS CHRIST

Jesus Christ our Saviour on earth, asks that prayers are sent to His Father in Heaven asking for His help and His love, so to treasure in our keeping for the sake of us all, for the sake of mankind.

Join me in saying a prayer, a prayer that can be said all through our world by as many as is possible, saying and beseeching our dear Father to have mercy and love on our blessed soul's.

Those of us who mean good, who wish to help and pass on to others thoughts of peace and balance be praised in the light of God. He has spoken to tell us the way to go beyond our normal thoughts and doings.

Take all that is happening throughout the world, those earthly events we cannot control without the word and interference of our Creator. Ask Him, think of sitting before Him and speaking in your language, He understands all. You are but a tiny part of Him given with such love.

Father in Heaven

Hear our prayer

Bring us forth to bear witness of your Divine Being.

We ask to be shown the way above all else in your eyes, to your house in the Heavenly vales.

Give us our thoughts to freely ask for your help with all that is Holy, seeking your help to free others from their chains of wickedness.

Give us the power to encompass love, spreading all to your children in need.

Help us dear loving Father to do what is right in your eyes.

We ask for forgiveness for the errors we have made, whether in ignorance or purposely meant, but just by asking for your forgiveness we may continue on our way, refreshed, thankful and cleansed of our discrepancies in your eyes.

9th April 2011

A Reminder

Tomorrow is a day of perfect love and very much to reflect on the happenings of today. If bad things have happened today, make amends for those things not to happen tomorrow.

We watch from above as the life story opens and the happenings unfold in their own evolution. We watch from above as soul's on earth seem to be able to understand that way of approaching where Heaven is, but we watch from everywhere, even a tiny distance from your face trying to encourage and perhaps turn the deliberations that are about to happen in the life's story, how it unfolds. It is not possible for us to interfere so when we see something that is wrong and that goes against our God's laws, it is known that we raise our voices in order to make a person hear our call of disappointment.

Your Guide has either a good earthly journey or bad with the soul He is mentoring going in the wrong direction or is led by another negative entity. God's laws are sacrosanct; they are all that we need to follow throughout life. Doing this will bring you to the end of your life's journey with proudness, vigour, setting our Father in your sight always, this is all that is asked of you.

Keeping God's laws in your sight always will deliver you to His sight with wonderment, having more knowledge from life's turmoil's. We know it is not an easy task, but rules, laws, were given helping all who follow with much endeavour to complete life's term to its full extent, letting our Father lead you from start to finish.

It is our joy to explain these rules on this page hoping that many souls, many persons accept fully this reminder of how to proceed from now onward. This is a reminder given with love because you do need to use these rules for your life's benefit, and as life can be turmoil and important things forgotten, here again is a reminder, please read and follow enabling us all to follow you with ease and interest.

God Bless You All

Words by Archangel Michael
3rd May 2011

Addressing a Congregation

Heaven is a place of love and Divine reflections.

Addressing a congregation is a lonely but loving lesson in giving our thoughts of love, reflections of our lives and future accomplishments, enough to start a healing process from words and prayers delivered by me today.

God is love and his words will not do anything to offend you.

The Spirit World and Angels guard our wonderful Lord and this is why all respect and rapport is so needed to show our devotion to our Creator.

Hallelujah and Reverence we sing and trust in all we learn and hear from Spirit's words.

All Is Beautiful

Words by Archangel Michael
12th December 2010

All Happened As Told

It is not easy to recognise where the beginning of a spiritual journey starts. Only by looking back over where so many happenings and wonderfully strange occurrences begin.

At a later stage in that journey when awakened spiritually, these deeds jump out and surprise you enabling light to shine surrounding the mind, soul and body with an unforgettable feeling of wonderment and awe.

My feelings and accomplishments brought out by this wondrous light, surrounds me with feelings of awe and the ability to go forward with every special thought, waiting with eagerness for more wonderful spiritual happenings. Now I know this is the truth, I know these are messages even signals from the Beings of our home in the world beyond ours where our loved ones reside, all waiting for us here on earth to see the light and feel that special love.

Remember when reading this book that all happened as told. Archangel Michael is my awe inspiring and beautiful Guardian Angel and Guide. I love and trust Him beyond thought and ego. He now is my way of life connecting me with our Lord our Father in Heaven.

This part of my journey as you will read started when I asked Archangel Michael if I could write a book of my experiences, enabling joy to enfold others reading it. I asked if He would help me write this book and then I asked 'Can I'. It was a wonder to behold the answer, who would have thought it giving such a complete and Divine answer to me. The answer to my question follows.

Will You Please Help Me Write A Book?

Carole you are able to write a book with my help that I am happy to give. The Angelic Realm and all who are therein will help dictate when you are ready.

The choice of prayers is decided by our Lord and Father He will take the interest. Think of how you would like to start this going. Prayers will be given explaining our home here in Heaven. Prayers will be given to you from our Lord on High, prayers of the highest order.

The book can originate from you but just by asking the right words will be forthcoming. God in Heaven has high values and will accept nothing less than perfect with his name behind any prayer.

You must start by putting together a beginning of your spiritual journey encompassing all the wondrous altercations you have been given and experienced through time. We here in the Angelic Realm have purposely and deliberately downloaded and sent to you these wondrous acts and experiences for you, to help you through and start your spiritual journey of which you are doing well.

The thoughts I pass on to you as you write are reminding you of notes written as a future reminder of these occurrences and experiences. At each location, each venture, each treasure you unfold will make a delicious insight into our wondrous world passing on your heartfelt experiences to those earthly souls who need extra simple and easy to behold happenings, enthralled into an easy to read easy to comprehend fascinating compilation of beautiful Heavenly and Angelic prayers.

God Be With You

Words by Archangel Michael
27th January 2011

Animals

You have asked Carole, for the next episode in the book. You have also asked what it could be, or will be, something that all will be interested in, something we have not mentioned before in these chapters.

The animal kingdom is a good place to start this journey. Firstly, mentioning the love we have for them and moreover the love they hold for us. What is a friend we ask ourselves? It is a soul who loves us, who would do anything for us. Love shows in the eyes, the window to the soul. Have you looked into the eyes of a loved one, an animal friend? Look and take in the very soul that is yours to love, to give a home to, to talk to, what an amazing thing to happen to take that responsibility on yourself.

Wild animals, sometimes tamed for our use, an example of this is a horse, now what a beautiful powerful animal to have as your own friend. What about the many wild animals who give to us so much interest to watch, to pet and stroke them for our pleasure. They were given to us all here on the earth plain for our own good from our Father in Heaven. How we use these beautiful animals is a test, a lesson, a learning curve for us to see how we treat them. They were put on this planet for us to care for them. They feel pain just as we do, have a heart just as we do, have children just as we do.

Some of us feel real fear for a particular animal, whereas others find love deep in their hearts. It is how we treat that animal whether we love or feel fear for it that is our learning curve. How many of us fear a serpent? Would you be able to walk away from it, leaving it to live its own life, or would you want to act differently? We are told they can only protect themselves by reacting as they do, whether we like it or not.

While we were holidaying in Spain we walked along a path to the local shop and could hear many birds singing it was so beautiful, we needed to stand on tip toes to peer over the gate. There were cages hanging from branches, doors, even a washing line. The little birds were so pretty and seemed so happy each one singing its own song.

At the centre of the small courtyard was a larger cage which housed a very colourful parrot. How clever was he, wolf whistling, copying everything people said as they passed the gate. Pass the gate they did, local people, young and old, visitors, each one spoke to the birds as they passed by, so why would the little birds not sing their colourful songs. How special were they, little did they know how they helped the passers-by with their songs of happiness lightening their day, also their little lives being lifted as so many people took an interest by talking to them constantly.

I asked Archangel Michael if he would explain the rules regarding the use of an animal for food. He said, "Sometimes over thinking is not a good thing to do. It isn't the end product of an animal turning to food, but the way it is done.

Humane is the way to go forward. Our Divine Father in His deliberations at the beginning of earthly life had to include a way of sustaining life, this being food. To raise cattle, poultry, all food supply must be done humanely; this is the test of our learning to love.

All these beloved animals are expected to be raised with dignity. When the time is right for slaughtering, this must be done with dignity and understanding. Any other way at all deviating from these ways are considered as cruelty. As you would expect this goes against the rules of humanity, love, light, also of life. Cruel ways only go against you at the last calling, it is not ignored."

Another Heavenly Message

Heavenly heart beats towards salvation and wonderment and mount up to a tumultuous level of vibration and beautiful endeavour toward enlightenment sent by God our Father.

Deliver Glorious messages of love. Bewilderment to all who suffer, knowing that God and a life await after earthly suffering. Give glorious thanks to our Creator for sending His Angels and helpers to sanctify the Cross and for giving us all Creation to love and work with.

Deliver spoken words of God's love and His patience and deliverance of all His souls, sons, daughters and memories of our home in Heaven.

Let love, light, deliverance, hope, fused with life's generation live in our very hearts.

Send forth forgiveness, life and sentience to all who suffer and have the need of our love, help and deliverance from bad thoughts.

All is not lost or over in this life. All will keep going forward as ever before with a new lease of souls giving more life to those in need.

God loves you always. He loves to hear your prayers even your stories of love and enrichment for when our earthly lives are over and it is time to come home to Heaven and our Father.

God bless you and keep you always in His sight now and forevermore my dear beloved children.

God Bless You

Words by Archangel Michael
2nd January 2011

Difficult Start to Life

It is without doubt a difficult start to life when the body is incomplete. The strength of the individuals' soul plays a major role on how a life will unfold, its play role given before the life starts.

An incomplete body is as beautiful to us here in the Spirit World as a complete one. It is how a soul handles the role given, to live as brave and as normal a life as is possible.

We love all on the earth plain, each has its own story of life to unveil helping all who come into contact with such a wonderful source of love and bravery, hopefully knowing that our Father, our Creator loves above all else those souls who are in need of earthly help. Not scorn, nor hilarity, but knowing there is a beautiful soul living a life inside that body, living the life as best it can for the love and knowing that a greater Being, our Father, is watching with awesome love, helping all to know how well they are doing, if not perfect by sight, but perfect in His eyes alone.

Send your help; by asking by prayer that the soul of an incomplete body is given the strength to live the time allotted in this His sight, until they return to perfect visual form in the Spirit World, Home and Heaven. So many incomplete bodies can so out play, out think and make a visually complete person feel incomplete and inadequate, so be it.

What a reunion at the end of that life there will be, how proud we all waiting will be and how proud of a job well done will that soul feel.

This is a very hard element of the learning that is required to take a soul forward in the Spirit World. It is by no means any less a responsible outing of a soul if at first try, it is unsuccessful. It is impossible not to learn to a certain level, even if not to be continued at that one time. Just feel proud to have agreed to partake with that difficult lesson.

You are loved beyond all else by your Father in Heaven and in His eyes you are whole.

God Bless The Meek For They Have More Strength Than We Know

Words by Archangel Michael
25th April 2011

Extreme World

God in Heaven gave us a beautiful world for us to live. He created the sun and moon, both of which give the opposite of extremes, freezing temperatures, great heat of which humans and animals are not able to tolerate.

Hunger is a solemn state and also to be without water in a drought. The world has amenities to cope with such things, some not fairly or evenly distributed throughout the planet. Even a desert has an oasis in the middle of all that sand to enable travellers to take in water, so to live. Think of the hot searing sun on the hot searing sand, but amongst this are nomadic people able to survive with their animals, camels, so named ships of the desert, they can carry water for themselves enabling them to live in such an extreme. This was well thought out, wasn't it?

Jungles so lush with colourful vegetation, plants and flowers, gloriously colourful birds so beautiful, yet so dangerous, with animals and reptiles living beneath and in that foliage. The rivers and streams that also house dangerous creatures as they pass through these hot, wet, jungles.

The world has such extremes of weather from freezing cold to heat. You would think that we could not stand the cold; there are people and animals that live in these regions. Polar bears for example, penguins, whales, fish and others, not to mention people who live in snow houses with their families. It is their home, it is where they live.

These extremes are brought to your attention to show how difficult it was to make our earth fit for us to live on, the sun giving light and warmth, with the moon bringing some moonlight and cold. We are also aware that the more warmer and exotic the part of the world the more the danger from earthquakes, tsunamis, horrific storms and other hazards.

We on the other hand who live in a mediocre climate, just hear about these terrible happenings from afar and can only watch and listen as these atrocities unfold. This is when the kindness of others' in the world comes forth offering help to the suffering and grieving and homeless, the voices together asking Our Father for His help and His love for those suffering.

Giving Words With Meaning

Yes, the words are wonderful, just of God from God giving words with meaning and words generating love that is so in need on the world of earth.

Generating the energy to accomplish feats of wisdom, knowledge and a generosity of mind is awesome in itself, needing your powers of thought to electrify energy for continuing with your light and your aura, giving the powers of learning here in Spirit the way of seeing your future purposes, energies and wonderful explanations of how our minds and bodies work and cope with life as a whole.

We here in the Angelic Realms report to our Lord all the possibilities rapporting to individual soul's putting them in scales of learning and knowledge gained as you live lives.

It is beyond doubt the most exhilarating experience to know you are the mentor of a soul doing a job well done. We feel so proud to be that mentor, that Guide, that Guardian Angel.

The souls who on a small scale gain their extra knowledge from experiences whilst on earth are cheered from us all from above. No soul needs to be a brainchild just the wanting to try and know God, to know Jesus, just to know love from the heart is the highest of mind knowledge and dedication that can be asked of you all.

God help you through your problems with love, welcoming you home when the time is right for you to leave the earths plain.

Words by Archangel Michael
12th June 2011

God's Wrath

We are at peace with our world until this day meets with the universe. It is the purpose of all who love and teach, leading us all through illness and bad hopes, bringing us beyond negativity and casting us through time and turmoil until the light opens beyond, giving light to all enabling the darkness to see beyond its blackness.

This blessing will overcome any deed that sits in the darkness feeding terror, feeding wars, feeding evil. Those who cause the growth of evil and do not see the error of their ways in time to repent and ask God's forgiveness, will receive all that is due at the life's ending.

There is no terror in God's name, no killing of innocents and no torture in His name. He is love beyond all reasoning, beyond all that He stands for as your Father.

Tell your dear Father in Heaven how much you love Him and how much you love on earth. Tell him what your input and participation is to help your world go forward in love and gratitude for all you are given.

It is those certain soul's on earth who cause such terrible atrocities in God's name, but how they mislead themselves, how they bring about the wrath of our dear loving, peaceful Creator.

Words by Archangel Michael
2nd May 2011

Heavenly Message

Dear Brethren and beautiful Beings, help each other to fend off fear and evil and learn about love and light, so that others may learn and communicate with each other and see and shine the light of God in our Heavens.

Tomorrow is another day; communication will be different in every way, depending on whosoever follows through with the teaching of love and sombrement with which we have to deal on this earthly world.

Beauty is all around, looking in wonderment at all you have been given to love and learn about our future home. Our home is Heaven.

Heaven on earth is a valuable place to behold, savour it and recognise it as your home with blessed thoughts of our Creator.

Archangels, Angels, Cherubim and Seraphim are around us as ever before. They guard our precious Lord and Keeper. It is up to us all here on the earth plain to do our bit in the earth's calling.

Welcome all your feelings with open arms and hands welcoming our Mighty Spirit in all we do

Love God always. Ask Him for help for others less fortunate than ourselves. This calling is a purpose built request for salvation and the epitome of strength in love, vibration and discourse.

Worship our Lord always with grace and favour, sending thoughts of love and density to all our brothers, sisters, all souls who require beautiful love and endurance.

Glory Be To God Our Father and Creator

Words by Archangel Michael
6th January 2011

99

It Is God's Wish

Cherubim and Seraphim surround our God in His world serving His every need. It is God's wish to be surrounded by creatures not of your knowing. He created them Himself in His eyes and His love, giving to them extreme Holiness, kindness, extreme knowledge and wisdom with an ability to be next to God serving Him only.

God is a power beyond any strength that you can imagine, therefore the need for trusting helpers to care for His needs is necessary to allow for His interest in the universe so vast, never ending, running our very existence.

There are many more Angelic Beings surrounding our Father each one further away from you, but nearer to His existence in the greatest Realm, the greatest world that is beyond your imagination.

Our dear Father is beyond your existence, beyond your reach. We are part of Him, we were born of Him and we must strive to return to Him. He wants this from you all, learning from your experiences through your adventures on a learning curve along your spiritual pathway of life serving Him with your love, your kindness and with fortitude.

The brilliance of His Godly brain has not been programmed into your earthly minds; to fathom His greatness is beyond your thinking. You are just His children learning about your Father who loves you beyond all else. His laws are sacrosanct it is not for you to question these laws just to carry them out as you are commanded. If this is so, your life's pathway will be clearer, your lives easier, your spiritual awareness giving you access nearer and sooner to your dear Father's world.

Words by Archangel Michael
14th November 2011

Just To Believe

It is without doubt that those people, those souls on earth who doubt there being a God or Jesus, or in fact anything at all to believe in lifting their lives, their spirits to the level needed to surround themselves with belief, honour and above all else the multitude of stories written for just those disbelieving people, disbelieving souls.

If only one person on earth is enlightened either by reading this book or talking with another who has, it can have a snowballing effect that enables another to begin their thinking program, leading to a small chink of light in the brain, that as thinking progresses the chink of light becomes a crack and so forth, leading to a miraculous enlightenment of another glorious soul.

This must have a knock-on process spreading the Word however small at the start, to many desperate, hopeless, unhappy, but mostly lonely souls who would love to be involved in God's light for His children, taking the loneliness away, giving life another meaning for them.

Now how wonderful is that, just one soul needs to have interest in the words of our book, to read this to family, friends, just spreading the beautiful meanings that are printed there, because they are not given from here on earth, but given by Archangel Michael on behalf of our Lord and Master, our Creator, our God.

All are words of truth, love and extreme Holiness that we must welcome with our hearts and minds open, ready to receive other Divine messages and blessings from our Divine Heavenly source. How can it be emphasised how important, how awesome it is to receive words of such glorious meaning, just sent for you to read and read again, to digest and follow through with so much excitement and conviction.

Where is evil, where is it from, where was it born? It seems that those seeking evil and following through with its extreme hate and unkindness, where there is no love attached to any feelings towards man nor beast, would benefit from feeling God's love and His mercy for those who regret their cruel lives in time to seek enlightenment, so to be helping others instead of causing immense pain and unhappiness to God's children here on earth.

God Help Us All To Believe, Words by Archangel Michael

Knowledge

The book is a beautiful venture for one so normal a soul. It is beyond all thought of education if that education has not included a journey to the Spirit World some call Heaven. Either word means the place you will call home, a beautiful, peaceful, joyous place.

Whatever you wished for when on earth, whether that be a pianist, any kind of musician, a singer, film star or maybe you wished for a large house with a garden. You may have wanted to be clever enlightening those who need your particular knowledge. Let it be as you wish, all is possible in your homeland of Spirit.

Whatever knowledge you have gained, however small an amount on the earth plain, all goes towards where you belong, on what wonderful level you will reside. No soul or Being is any better thought of in Spirit than another. The difference being the amount of knowledge you allowed yourself to know.

Green fields, trees, plants and flowers are in abundance here. Yes, we do get rain just enough to water all; we are certainly not awash with rain as you can be on earth. Your home in Spirit World is a wonderful place to be. There are no time restrictions here just the beauty of all you want to be, see and have.

We, all who are involved with you, know when it is time for you to come home. It is all given telepathically so your loved ones, even animal loved ones, won't be late for your wonderful exciting arrival. We all are aware what you have gone through experience wise, so to crowd around you upon your return is just a miraculous experience not just for you but us all here at home also.

Those dear souls who have lived a long life and are too weary straight away to celebrate are helped to rest, refresh themselves and their energy, before they take their natural place they so deserve in Heaven.

There are darker places for those souls who do not want to see the light because they have lived a much lesser a loving life on earth due to acts of cruelty to others. The word cruelty covers many reasons, abuse, and warring.

Head for the light always my children. We love you and await your return with eagerness.

God Bless You All

Words by Archangel Michael
24th February 2011

Life Goes On

Spiritual life is the essence for our soul. We wait for our loved ones with love, waiting for their term to finish on the earth plain. So many souls' can relate to an earthly life because they have been there, some over and over again.

The kindness and love of souls here in Spirit World is so exacting that even when they watch and wait they are sending you those feelings and joy. If you talk to your friends and loved ones in Spirit they do hear your messages, even answering you when they can. This is not always possible to get through to you, if you are not receptive to their vibrations. Maybe you are not sure, maybe you think you are talking to yourself, well you are not and we know that is not the case.

Life goes on; you are the same personality as on earth, the same personality you were before you left home for your first earthly visit. Your loved ones are kept informed as to how you are doing with your journey, they are also kept informed when you are arriving back home.

In the Spirit World you can be where you imagine you would like to be, the seaside perhaps, walking through a forest, climbing a mountain, riding a horse, it is your choice, all is possible.

Knowledge is the way forward for you to reach higher possibilities, teaching your soul to learn and be more spiritually knowing. Learning more about God our Father and how you can let Him know how well you are doing in His eyes.

It is known amongst us all how difficult it is to live on planet earth. We watch in amazement as lives and stories unfold. Are you doing well or just OK? It is whether you try to make your journey in truth or not. Love life; ask to be shown the light, increasing your wellbeing and success through learning as you go.

The Bible

The Bible is a classic book telling the world of God and Jesus in easy to understand stories. The Book was given to us by our Father Himself, telling us all there is to know about Himself and our wondrous home in His Heavenly place of worship. There really is no need to read any other book about Godly history and happenings, if at first you read His Bible.

The Ten Commandments were given to us to show us in His words how we are to live, just living by His laws, setting those laws in easy to understand rules laid out in simple sentences for all to read whether scholarly or not.

However a soul starts out its life, there has to be rules for a way of living, not just floundering without knowing a way forward in order to keep our world a fit place for all to live.

If you are to live a happy and contented life giving off love along your way, giving help to others, then how helpful would it be easing your burden, knowing what is expected of you. It is good to know that rules exist, how many without those rules would murder, rape, steal, if the words were not written in stone telling us all that these things were not expected of us, that they are against God's laws laid down by no other than Himself our Divine Lord, and written in the Holy Bible for us all to see and read by Him.

He created the universe, he created our world and all the wonders therein, it would go against everything just to leave us to our own rules and imagination and ruin everything for ourselves and everything in it.

He is the one who sets these laws for us to obey here on the earth plain; they were here at the beginning of time. We are asked and expected to follow with all our hearts and minds following with love and devotion with all that is asked of us by these commandments. As the word suggests they are not a request but a Command from our Creator for all our benefit, to bring us through our lives by following what we have been told to do by the Commandments and the Bible.

Words by Archangel Michael
17th April 2011

Today Is Another Day

Today is another day in our lives to celebrate our earthly knowing. Without that knowing all would seem an endless task of thought, trying to mend what is searching more and more for an experience that never ends.

What would we as individuals prefer: to pass over to our home in Spirit or for death to be the end? Spirit World is where we originate from, it is our home, and it is astoundingly beautiful. It is for us to see and for us to know and believe in. Surround yourself with this belief and knowledge. What a wonderful feeling to know and believe enough to pass on this belief to another unsuspecting soul on earth, who has no idea what to think and believe.

Jesus Christ was sent to earth to set us asking about God in Heaven, miracles, wondrous occurrences and how our beautiful Saviour suffered for us in the name of his Father, our Father. Would we be thinking now if it was not for Him?

It is beyond our comprehension the beauty, the Holiness of our home in Heaven, in Spirit World. You are asked to think and take in to your mind all experiences told and written. These words are exactly from God our Father in Heaven who loves you above all and will welcome you home when your time is ready. You are asked not to be frightened but believe what you are being given. The words are truth. All of you on the earth plain have the ability to read and believe in others' experiences and this is what you are asked to do - believe.

To Trust

Spirit must in itself be special and endearing. These words must encourage you to ask why, is this real, is this true, can I myself have some of this spirituality. Well, the answer from above is sincerely 'Yes' and encourages you to seek meaningful messages which encompasses you and gets you to question what you see and read.

Trust must have a lot to do with this. Do you trust in what you see? Do you trust in what you read? Read these coming prayers, they are for you from our Divine Father to give you faith, peace, and wisdom.

Definitions of God our Father surround us with stories and prayers. Prayers of everlasting love making it more difficult for us here on earth to believe what we hear, even though we seek to believe the wondrous stories that are Heavenly sent. Wouldn't it be wonderful if we all could believe and know of our home beyond the horizon? Would we enjoy our lives much more? Would we when in troublesome times find it easy just to end our life to go home? This is why our beautiful memories are taken away from our vision from our memories, to encourage us to live our lives individually and see our journey time that has been allocated to us. We must try with all our hearts and might to complete this journeying time.

How are we to listen and then understand the word of our Father without first shedding that beautiful light, enabling us to prosper and go forward with all the love we can muster? The idea of our Father is to give these words from Him only to us all on the earth plain. Enjoy and trust in His words as they are truth.

The words in this book portray all that is truth in Heaven and love beyond the clouds, the stars, and the universe. God is asking you to understand that He is all around you. He and all in Spirit in the Angelic Realm in all Realms are only a touch away.

The energy from God is so bright, so strong, that ordinary mortals are not able to be near to this magnificent Holy Realm that is called Gods House, His home, and His beautiful Holy Realm. His home has flowers beyond our imagination, colour we are unable to see or know. He is surrounded by miraculous Beings all born from His light.

Waiting To Start a New Life

It is with knowing God that events become evident, that those of us who believe and learn about our dear Father, survive to show others that love is a beautiful thing. What is wrong with learning to love, what is wrong with thanking our Holy Father for the wondrous events we have been blessed with?

Standing and waiting for a foetus to become available to take upon it a new soul is a nervous situation in itself, but once all is ready, and the right one is found, the life's journey is ready for new life. It is a long process from conception to rebirth that the waiting soul may continue with its usual accomplishments, returning when the time is right to the waiting unborn child.

This is a new life, this is a remarkable event. The soul who has taken on this highly wondrous event must stand by the decision made and commit to starting out on this venture, and that is to live a new life with all its heartache, worry, grief and leading up to those feelings, love. It is love that decides how you succeed or not on this course of events.

Knowing God, respecting the Guide who you have chosen for yourself before using the agreement of birth and life is one, if not all, the secret of a successful time on earth, hopefully helping others. Forgetting the names, reasons and thoughts that you knew and were aware of in Heaven must give you an enormous lift. Remember you are not meant to remember from whence you came, the soul's you knew and the soul's you loved and really did not want to leave.

In spirit you as a soul are extreme energy. You do not use that energy all at once, but take only as much with you to the earth plain, enough to help you through the roughness and softness of living. The remainder of energy stays in Spirit waiting your return when earth life is complete. Think about this, because you must be there with all the other relatives and loved ones. Does this mean we should not grieve for our newly departed loved ones as we are already with them in Spirit?

What a beautiful thought, but a very difficult one to behold. How we must let ourselves believe by taking in and using knowledge for us to become spiritually active here on earth.

Joy Be To All, Words by Archangel Michael, 17th March 2011

Wonderment Is To Behold

Wonderment is to behold our beautiful world with all it contains. Flora and Forna: birds of so many varieties, animals wanting our love. Each beauty you behold that you look at, wonder at, even to take advantage of or take for granted is planted there for you all dear soul's to take as your own to love and deliberate over giving your thought, love, beauty and all to call your own.

You are not alone; you always have a belonging to call your own. Look and see, learn and take as your own and with kindness, show these wondrous things to enlighten another's life who has not yet seen the beauty which is theirs. All is given for all on this beautiful earth.

Look, see, listen and take in the amazement of all that is given you in our Father's name.

A Wonder to Behold

Words by Archangel Michael

Words For Those Who Are Sick And Grieving

Divine souls of the earth and universe deliver these words to the sick, grieving, frightened and to those dear souls who would ask for courage and strength to carry forward their feelings of death.

Death is not a word we here in Spirit like to use because the word death means the end and the end is far from the feelings we should be feeling as we are ready to pass over into our wondrous World of Spirit we all must call home.

It is difficult for you to think and imagine that word death. You will pass to home without stress or pain. It will be beautiful for you to arrive home. Then you will wonder why you did not believe and cross over the divide with ease and peace.

Waiting for you will be spirit's, soul's, loved ones, friends including your much loved dear animal friends who await your return just as much as a human friend.

Tell all, speak all about how you feel even if frightened and speak about how you feel. Talk to relatives and friends but most importantly speak to your Guide. Do not think, I don't know, I don't believe. You will be doing a great wondrous service to yourself just talking, asking questions.

Crossing over to your home in Spirit is the easy thing to do. It is the frightened part that you must cope with before this happens. You are loved and honoured, spoken with all our hearts. Listen; believe though you may find this difficult to do. We are here to welcome you. We are waiting to give you peace of mind, any assistance you need when you arrive. Love, listen and believe all that is given you as it is the truth.

Bless you who are suffering. Bless you who have lived long and who are tired of this earthly life. Bless you all in God's name.

Words by Archangel Michael

Your Guide Is With You Always

Every day is an opening for us all to feel joy, opening our hearts to all who suffer and do not have any idea of asking our dear Lord for help. Each person or soul on earth has a Guide who looks after that individual person, showering that person with help along their life path. How that person or soul responds to that Guide depends on how much they are aware of His existence.

Although some may not know it, their Guide is the most important being there to help, if help is needed, because that Guide is your very best friend if only you knew about Him. It is only by following your own beliefs and allowing others experiences to penetrate your mind, will this interest allow you to evaluate what you have learned gaining this glorious knowledge, this Being, the reason for your existence on earth.

Think about your future and what it would mean to you immediately, but most importantly, what it would mean to you and your close loved ones, for you to encourage them with the learning's you yourself have taken the plunge to know.

Your Guide has been assigned to you from birth and is there to nudge you in the right direction if maybe you are ready to take the wrong road at a crossroads. So, you will be given a nudge, a sign that you may take notice of, or not. How helpful is it to know in your mind that you have that Guide and the more you know the closer He will be to you knowing you are ready for a sign.

Your Guide may be a soul who has experience of living on our earth plain, or your Guide could be an Angelic Being, an Angel. Wherever this Guide is from, which plain of existence they originate, their hope for you, their dream for you is set deeply in to their existence and their reality for each of you is love and learning, giving you the knowledge to take back home when the time is right to our Heavenly Father, who asks that your knowledge takes you closer to Him.

Glory Be To God on High

Words by Archangel Michael

More Words of Meaning for Us All

~*~

A Perfect Rose

It is without doubt the most beautiful world that the souls who are asked to reside there can experience, as they are given the thoughts and wonderful adventures as they continue with life's plan for living a meaningful existence, whether that be hard or easily lived, the world they are given to live on is somewhere to be proud with its beauty a feature of any existence.

The difference between beauty and ugliness, hatred and love, hot and cold, light and dark, man, woman, are all there for a reason, that you are given the choice on your beautiful world to follow beauty or ugliness, to feel pain or the strength to endure this without knowing the extremes.

The earth world is a place of delightful pleasures of seeing, hearing, experiencing all that has been given, enough to give you knowledge on how to love and care for all humanity residing on this planet. All that is living has the perfect right to be there, therefore, it is in your grasp to let the life there continue growing and increasing in size, content for you to watch a thing of beauty evolve.

To see and to smell the perfection of a beautiful rose in a garden to give you pleasure, the perfect scent of something oh so beautiful to give you perfect peace evening up the feelings of turmoil and tiredness, each feeling out weighing the other.

It is noticing the opposites that can help you appreciate each side helping you to see what is around you, letting you make the choice of good or bad enriching your living opportunities to encompass what you think should happen, not leaving it to another soul to maybe make a bad choice on your behalf.

As is a rose opening its flower, its petals giving off the sweetest most beautiful scent from such a small but extreme beauty that the earth can offer to man. All things are made beautiful for the earth plain to be so proud, the earth word meaning that which allows something so delicate, its roots finding sustenance and water from it, to enable its existence just by extracting warmth from the sun, giving it life as it does to all living creatures except those of complete opposite existing at the depths of the ocean where crustaceans live in complete blackness.

Carole M. Day

Search for beauty in all things as easily as searching for the light of God, the beauty is there to be found. When you find this light, this beauty, however small a chink of light you see, hold on to the sight of this wonderful scene, keep this with you always encouraging a smile to bring a happiness, uplifting your being from negativity, giving this smile, this happiness to another human being in need of enlightenment.

Words by Archangel Michael
29th June 2011

All Things Beautiful Are God Sent

It is without doubt that all things beautiful are God sent to you all on earth with so much love. We here in Spirit and Angelic Realms study the way forward for you all with might and Heavenly love.

Your lives lived at the moment of our communication are looked upon with esteem and awesome wonder, because we are so aware of the strife and difficulties that you face as you journey through your lives. You are sent to the earth plain by soul's who are wise with wisdom knowing to where you are being sent, not letting you be anywhere that does not originate from your Creator. You are sent with the knowledge of great success being in your auric field.

You originate from love, you originate from a loving Father always showing His children the way forward, hoping beyond all else that they are knowing of their birth place whence they came, giving courage when faced with bad and evil, recalling the wonderment of their original place called home.

To leave the Spirit World called home with this knowledge that you are surrounded by wonderful Beings, the most awesome and wonderful Being of Beings is God our Father who has given us life, has given everything we have in the name of love.

Listen to all that is yours to behold, believing that you are not alone but always accompanied by a loving, caring Father who has given to you a faithful follower for you to talk to and know is there beyond all other friends. Your Guide is with you always as a gift from God as an all knowing friend, there solely for you. It is in your interest to feel this with all your being for your own true living of a beautiful life.

God Be With You Always

Words by Archangel Michael
9th July 2011

A Mind Blocked With Thoughts

For a week or two in August 2011 I found it difficult to receive the words normally given to me. The more concerned I became, the more I thought my Guardian Angel and Guide, Archangel Michael for some reason had stopped sending to me those beautiful words and prayers. It didn't make sense as I had been told that I would receive help to write book two, so what was happening? I asked that question and received the following answer.

I could not leave you Carole, I will never leave you. It is my special wish that I give to you messages of joy of great help and interest to you all there on the earth plain.

There are many Angelic Beings and those in the Spiritual Realms that have such a lot to say to tell the earthly world, giving their stories of encouragement helping you along on your journey. We are able to watch everything that happens with your lives. We can see errors that are made before they actually happen; it is a life's spiral.

Sometimes, if you have not taken the care with relaxation, if you are hurried, it may not be possible for a free flowing message to get through past your thought pattern that is so blocked with thoughts, too busy to let our light flow through.

This morning at your development circle it was so easy to send through thoughts of messages because the relaxation of your mind was so successful enabling thoughts of messages to so easily get through to your mind slant.

Generously seeking messages for others is for you a wondrous event, to us, we are just waiting for the mind development to be sufficiently opened allowing us to encroach your thought pattern. For us, it is a natural way of existing through mind thought, but for those on earth it is a hard task to push through a mind that is blocked.

The earthly way to doubt yourself, with that doubt a shield is put up between you and Spirit. It is only when that guard has softened does it allow contact between our two minds and our two worlds. It is an experience to hold on to because once understood, brings us both closer to each other allowing a contact that is so wonderful to share, which is a real development of the minds which will continue expanding over time.

It makes sense when explained that to make contact with another in the Spirit World takes more than just sitting there thinking that it should happen. What a wonderful experience to ask God to enable you to speak to the Spirit World to access the light that is so important in contacting another world of existence even though we know that world exists, oh it does exist. If you believe what a wonderful thing to value and know that your home is there, your loved ones are there to speak with just by asking.

You are seeing beyond the normal Realms of existence, you are seeing the light even though not as extreme as contacting the Higher Realms where our Lord Jesus exists, nor where the Archangels and Angels exist, nor where our God on the highest level of levels exists. That which takes extreme time, time that is not possible to achieve only with learning and gaining those heights of existence that is asked of us all as we further our lives on the earth plain starting us on our journey so wonderful, fulfilling our destinies over time, time that does not exist in our World of Spiritual fulfilment.

Words by Archangel Michael
14th September 2011

Another's Viewpoint

It is only with great patience that all on earth may live their lives in entirety slowly working out what has been asked of them. To know without incurring any problems is a wonderful thing because knowing the right thing to do, the right way to go, is given to a soul before birth. Having this knowledge and following a pathway easily, is more than is expected of you, so to struggle with the path given to you should be worked out by the soul and that souls Guide assisting with your passage of time and existence.

Just imagine sitting watching a film and as part of the audience you can follow the story giving an opinion, thoughts as to whether the plot or plan would have been better following a different direction, giving the end a stronger and more discerning ending.

That is how a Guide can be likened to surveying your story live on the earth plain, knowing that had things been enacted differently the outcome could have been improved. Life is a wonderful adventure, sometimes the drawing to an end of that adventure is not as was planned but the soul can feel proud of the input looking at it from a different angle.

Think of your lives as an artist views his work, with excitement putting into his work what he thinks life should be, how it should be captured on canvas for not only himself to see, but all the world, if the world holds the same excitement wishing to see another's vision.

Give yourselves this viewpoint to help see beauty, or pain, instead of seeing nothing but your own thoughts on the journey of lives. There are many people on earth who have the capacity to expand their minds, enabling them to look upon the story of evolution, the wonders of the universe, putting this enlightenment on to canvas, to music, to studying and passing on the knowledge to another who is hungry to receive the information to further his education and experiences.

If you feel that you have a gift from Spirit and this has remained unnoticed so far, spread your wings and show resolve by giving that gift to others to see, feel, read and experience encouraging your light to shine in abundance into another's life that is suffering by a temporary dimness, just waiting to be uncovered by God's light, giving the necessary strength to go about your planned journeying.

You all need the help from a friend or a loved one at some stage of your journey.

Words by Archangel Michael
26th August 2011

A World of Thought

Wondering through the world of Spirit is a beautiful and idealistic thing to do. To go where you wish to go in just a thought, to say just what you want to say without sound, just another thought inspired way of communicating.

To hold the idea of the wish to be in a different place of existence, to see and experience another place of beauty in just the quickness of a thought is an enlightenment of experience without the heaviness experienced on earth.

All here in Spirit is just a thought provoking world, the wonder of realising how to live a life without the turmoil of a noisy and sometimes violent existence, to realise the choice of quietness and peace if so required.

We know that when you return to this world of experience you will be back where you belong in your home of existence before your earthly venture originated. You are able to visit friends, loved ones with just a thought. It is possible to visit different worlds, live lives on these worlds if you so wish it.

What a wonderful thing to be able to achieve pursuits you had wished you had carried out previously, or knowing in hindsight what a wonderful accomplishment it would have been to sing, play an instrument, to do anything not in your powers of learning whilst on earth. If you would have loved to play a piano, be a concert pianist, dance, or maybe to help another person on your earth world in need of your love, but then did not have the chance to fulfil this objective, you can with just a thought here in the World of Spirit.

There are many soul's here putting their experiences into book form of their lives on earth, each giving an account of a completely different way of living in a different part of the world of earth. It is a pursuit that gives not only delight to the reader of the book in Spirit but to the writer who is personalising memories and carrying out a life's play putting action onto paper.

All souls' completed lives lived whether in a positive way or negative way, are captured forever and filed in archives of lives lived from birth to a finished life however short that may have been, allowing that life to be evaluated by Beings of extreme wisdom and higher powers of teaching, their guides and for the souls themselves to judge what scale they have achieved throughout their earth lives.

What wonderful words to just appear out of nowhere, or so it would seem. As usually happens I asked who it was that gave these beautiful words through to me, and the answer was:

"It is Miranda sending through these words of simplicity, these words of truth as I see them.

The actions from you all on earth are so difficult to define by me here, all is so extreme, which means extreme feelings that perhaps I could not cope with if I were in the same place experiencing the same things.

Love me as I love you as I journey near to your life's conquests. I wish for your blessings throughout your journeying on earth, you being nearer to my world than you could possibly imagine.

God bless you in all you succeed in doing throughout your life's ventures"

Miranda
11th October 2011

Come Forward With Your Thoughts

Now is the time to come forward with your thoughts of wisdom, thoughts you have accumulated in your mind about our beautiful spiritual stories, enveloping our very soul, entering into a world of love and belief in a home beyond earth, just waiting for you to arrive back to the place you originated.

There are many on the earth plain who teach their ways, giving wonderful experiences and collecting many vibrations on this earth, there meaning given with such honesty, giving their experiences to the excited few who would listen with minds open and an interested volition, just taking in the awesome words expressing life beyond the veil, with extreme belief the audience taking every word of expression explaining about our Father and those extreme Beings who wish for your love, kindness, belief beyond all else, just wanting your belief in their Holy Realms.

There are ordinary everyday soul's here on earth now, who give their time, knowing how we grasp the interest as they give their talks and explain the way forward as they see it and know it, just to attract non-believers to their knowing, giving them a reason to fight for their reason to live and reside on the surface of this wonderful world. It is telling them of the awesome home that is theirs after finishing their time on earth.

Those of you who have the belief, the telling, the love, follow through with your thoughts and express yourselves enabling more news of our Father to penetrate the land. Seek and ye shall find, is as God asked Jesus to pass on to us all. If you have the interest to seek, it is more likely that you will find, when you have found, how beautiful is that to add to your cherished dreams, memories and thoughts of that momentous day.

I am an Angel of Christ and with gratitude to Him ask that these words be included to form a manuscript of true passion of the completeness of our homes with our Father and His Son Jesus Christ.

God help you to have the strength to succeed with your lives knowing you are looked upon with love all the while you are on the earth plain.

God Be With You

Words by an Angel of Christ
7th August 2011

When writing the words "*I am an Angel of Christ*" I wondered if this Angel had a name that I could add beneath His words. I asked that question of Him, do you have a name?

I am an Angel Of Christ, with all I am I serve my Master, wishing for no other than to have my words included in the manuscript to be circulated amongst many souls in need on the earth plain.

God is love.

7ᵗʰ August 2011

Gently Is the Way Forward

Gently is the way forward for those souls who are struggling with their life's path the journey planned out at the beginning of their earthly birth. Look forward with persuasion of doing the right thing with the next stage of your journey.

Think strength before any other thoughts. With strength you can conquer all, otherwise there is only weakness. Fight your cause and arrive at the next scene with vibrancy and fortitude because without that life will be more than difficult for you which is an unnecessary experience.

The life you have chosen is strewn with many crossroads with many different paths that you could chose to follow. The pathway you take on that life's journey will be an exciting expedition of so many experiences.

It is not possible for any soul on earth to know where they originate. It is only by taking in the feelings that go along with the mistakes made as we journey along our road. We alongside your very soul travel our journey of our calling hoping above all else that where we are going is illuminated along the pathway, giving an insight into all that will take place along our route of ventures.

If you take stock of all you have done over time, then take stock of what you want to do in the future it should be an easier burden to carry on your journey, lightening the darkness of unknowing, lifting the veil that is covering all that we are and all that we could be.

Fighting for a decent life amid all the feelings felt in reality is something that only a job well done can accompany us all in life's term. Take stock of all you do, letting those who are interested into your feelings of spiritual awareness knowing that you are facing in the right direction looking in complete knowing of where you are and where you will be in the future.

Light up your lives, look to the beautiful light where you can face your God, letting the darkness fade into the distance, your mind occupying the awareness of awesome wonder as you get nearer to your Father.

God Is Light

Words by Archangel Michael
22nd August 2011

Greatness Is a Possibility of All Men

Not one word has been added, taken out or altered in any way from the following wording.

After scribbling down so quickly the information I was receiving I wondered if the words were in any sought of order as I found it difficult to understand and to punctuate the content, needing to read it over and over again.

Having read the document many times the meaning has now become apparent to me and I take great delight reading these wonderful words given to us all from another dimension.

~✳~

Greatness is a possibility of all men if they seek solitude and tranquil abandonment, settling minds to a slow pace of peacefulness.

It needs peace and solitude to train the mind to receive greatness referring to all the mind has trained itself to take in, that being Divine information given from the higher source of knowledge and wisdom whose only reason for passing this knowledge on to you all on earth, that is to ascertain whether your minds are able to receive this awesome knowledge that is such an exciting feeling to experience and being able to accept this information, God's knowledge.

Not just accepting what you have and not which you could have, but taking in all that is possible enabling you to achieve that greatness sought after but not always obtainable by your own means of learning.

It is only with such completeness of excessive information given to you while you are relaxing or meditating is it an easy experience to achieve. A brain is a glorious achievement never used to its extreme advantage only a small proportion is used at any one time.

Use this time of enlightenment to Glorify your Creator by using the brilliance left to you to reach every corner of learning, giving your brain a reason to receive information from our Divine Beings just wanting you to achieve all you can by taking in that knowledge and wisdom that you did not know was there for the brain to store, encouraging that light of brilliance to shine throughout each demeanour creating that greatness that is for all to maximise to full strength.

This message these words, are sent to you by Me here in my Realm of Beauty and Love.

God Bless All Those Who Read These Exalted Words.

Words by Archangel Michael
6th October 2011

If Not Him Then Who, Most of All How

Whatever reason that you think you have in not believing in the wonders of life and life after death, you are cheating yourselves missing out on a multitude of glorious aspects of brilliant lights welcoming all to the wondrous world of your spiritual link.

We wish here as we watch the acts of your stage play that you are aware of all we are aware of as your lives pass by, as your earth hours tick away the very minutes encased in your journey of belief or disbelief the choice is surely yours.

When you return home and brave the lights, the honours, the friends and loved ones just questioning you with *'why did you not know that you had a Creator? Why did you not know that there has to be more to the universe, to your world, to your way of living your life?* Just by thinking about everything that makes us who we are, then just think "*if not Him, then who and most of all how*". Just by knowing that all is wonderment and doing the simple thing of showing our God that you believed in Him before anything else'.

Solving the great conundrum here on earth of how to journey through space at these incredible speeds can only tell you that the lives you are experiencing now on your planet are but a fleeting thought in the never ending purpose of the universe and all this created by the Creator of all things.

If you as inhabitants of earth take time to think of the wonders of living, the wonder of your earthly home surrounding the small part of the universe you can see with the naked eye, the nearest part of this universe for you the sun and the moon that can seem so close on a clear night sky. When you can see beautiful planets quite near to you then imagine the travel possibilities of the future to worlds in far off reaches of the universe, meeting with other life forms and learning from them, perhaps learning how to cure disease from those life forms who are many millions of your earth years' of life lived, giving an excellence in experiencing knowing and loving as you are asked to achieve here on earth.

You are not to know how these life forms from beyond our spectrum, millions of your earth miles away, how they do exist. You would hope that evolving through time would stop wars, dictators, disease, letting life continue its way forward just helping those worlds not so advanced to see a clearer picture.

It is true that life forms and the way they evolve cannot be tampered with, cannot have their futures altered in any way by another thought form, it is up to those inhabitants of their own worlds to make a success or not of their own destinies.

25ᵗʰ November 2011

Journey to Earth

In conversation the question has often been asked, *"do we know how our lives will end before we actually leave our homes in the Spirit World to make that journey to live a life on earth? "*

Your journeying to earth is a wondrous occasion here in Spirit. All things are equal amongst soul's, it is only by request that a soul is invited to make this journey to earth for whatever the reason for the visit.

It could be that a life was not fully completed, cut short by the soul himself, making the time they did have not a long enough service to gain the knowledge and experience required to bring them a stage closer to God.

Yes, is the answer to your question, all soul's visiting earth are fully aware of the expectancy of life, how it will end, by whatever the circumstance. Those souls who fall at the first hurdle because of not following the pathway offered to them before birth can pick themselves up and continue along their way learning from the experience, it is the plan.

Others' will mix with soul's who have bad or evil tendencies instilled into their brains before a soul visit's the body and are not able to help themselves bring their journey to the right pathway as was previously designed in the plan of life. God is all seeing, all knowing, able to see to the future of all His children.

A soul may return from his trials on earth quite sad and disappointed after going over the mistakes that were made on his visit, perhaps a few simple mistakes or possibly making a complete hash of it all. What should have been an interesting viewing of all life on the earth plain, corresponding with the arrangement, following through a laid out plan of existence but without success. It is then for some, so very important that they request they go back to earth for another try at the trials and tribulations of living in such a harsh environment as Earth.

The whole plan of life on earth is to learn, to experience, to gain that unfaltering need for knowledge to be taken back home with pride to enable that soul to further his existence to be nearer to our Father.

For some the need for experience and knowledge of one particular field of feelings, maybe to know the pain of grief, or loss of a loved one, or intense love, maybe needing to feel the pangs of jealousy, it is all there on the earth plain ready to give to you that lesson you have requested out of life, for you to cope with that intense feeling.

To take one's own life ending what should be a beautiful experience for the soul involved is a sad occurrence. We know of course that the human mind is exactly that, human, sometimes it is not possible for the human mind to cope with all the worries of living a life, there being a weakness in some, giving their life a more extenuating reason for taking another route out of the difficulty.

We are sad here in Spirit, but more than any other we are understanding of the plight of another soul in difficulty, this is a further reason to take another trip to earth to try again with all the given experiences, this is the soul's choice. God is so proud of the efforts made by His children however many times it takes for success; He is an understanding, loving God.

Words by Archangel Michael
4th July 2011

Knowledge So Vast

It is with these words that I encourage you to think and deliver your thoughts to those in need of your story telling, the words written with love. Your words written with others in mind are to pass on knowledge from another source of writing.

Whilst others wait with excitement of experiences from another world, words giving you an insight into another way of living, most on earth live to breath thoughts of the unbelievable, a difficult process of the mind to imagine another race, another world, another universe, but if you have the belief, if you have an open mind, the excitement opening the mind enables the outgoing thoughts of wonders beyond our thinking.

To those who are able to expand their senses allowing another way of life to filter through into their thought field, normally locked due to their own involvement in blocking a wonderful mind from knowing the adventures of the future to be shown, giving a glimpse of the impossible and turning full circle to the possible.

A mind should be open to the wonders of the infinite universe with knowing and knowledge so vast it is difficult to go further to believe more than the edge of the precipice, ready to begin the process of unloading its wonders to those seeking the answer to that universe.

The information deemed as knowledge is so vast that only a small amount at a time can be released to keep the pressure at a comparable level for us all to take, stopping an overload of information to crowd a mind at one try.

Words by an unknown source
15th October 2011

Profound Love, Love Is Light

For some on the earth plain religion is frowned upon not to say the least ignored, just not thought of only in discussions.

It is here in the Spirit World, your home, that all feeling is religion, but we have no need to think it so because God is love, meaning that all; everything to the light is profound love. Love is light, light is stronger and brighter as you are nearer to our Holy Father.

The energies are beyond your thought, but we here learning our course and how soon we might go forward in beautiful excess of learning of our Creator in glorious demeanour, showing our education by colour beyond description and giving insight into the levels of how much energy has been taken into a mind to enrich and glorify the complete energy our Creator has to offer us all in love and glorification, enriching us one and all giving to all our souls who glorify Him the encouragement to take a step closer to His magnificent Kingdom.

It is with that step that we look to and encourage ourselves to take using His name to further our ambitions, to thank our Father for our home of beauty, trust and love beyond any other knowledge.

Look Beyond to the Light, There Is Love

14th January 2012

Sensations

All summer long on the earth plain the warmth of your sun gives an insight into the way you are made to feel, how light hearted, how well you feel after the long winter months have subsided leaving you with the wonders of the sun on your body.

Here in the World of Spirit your home, you may experience whatever sensation you are comfortable with, or be where you would like to be, the choice is yours to experience. There are magnificent gardens with plants, flowers, trees of colours you could not dream, there are snow-capped mountains, if that is what you want. Close your eyes try and imagine many other different colours, many other different flowers, just a thing of utter beauty.

It is not for you to know, for you to experience the spiritual beauty that evades you until you return home where you can experience the many wondrous things of pure beauty, of unimaginable shapes, sizes, colours. The beauty on earth is in itself marvellous but not in accordance with your homeland waiting for you after your sojourn to this planet of many difficult dreams.

Distinctly positive vibrations are here in Spirit where negativity does not exist, instead of hearing you cannot do that, it is in an instant thought that you see, feel, hear and simply travel to your far off lands in just a thought.

To dream is another temple in your journey of lights. Thoughts in this never ending desire to experience your life originally lived in your world of Spirit, in your God given place of existence of love and kindness, knowing of deep feelings of Holiness as you seek that beautiful light leading to those dreams allowed by your Father, a glorious founder of your home, a universe of splendour given to you for your own happiness and fortunes of faith following our learning's through time, lighting our pathway of Heavenly wonder.

Greetings dear people of our world grant us the knowledge of your earthly deeds; we await your wondrous return with excitement. Allow us to welcome you with complete love wiping away the pressures of a life lived on such an extreme planet. Well done.

12th December 2011

Some Teach That It Is Evil to Make Contact With Spirit

Death is a wonderful thing to experience if only you knew it at the time. All it is, is your life term on earth coming to an end and that happens to you all, to everyone that live their lives going back home when the time is right.

Our Father asks that all soul's on earth believe in Him and what He has to offer, which is retrieving His children and bringing them back home to Him, ready to find their way to the part of Spirit that has been given to correspond with what they have learned, where they will continue learning taking them nearer to Him.

If you do as our Father teaches, if you believe in Heaven and a home there waiting for you, how can evil be at all relevant? If all you are doing is contacting your loved ones who have gone home there can be no evil in that.

Is it right just to sit and wait through all the days of your life just waiting to see what, if anything, happens? No, you are seeking God; you are seeking your home. God gave you freedom to do with your life what you think is right, to believe that God, love and Heaven is right, just reach out to touch it, to find it. We are here hoping that is what you will do. There is no evil in this; we await your deliberations, your thoughts, your trials.

It is only belief that encourages you to talk to Spirit, to your loved ones to your Guide. If you don't believe then there is nothing for you to search for but to just wait for death at the end of your life.

God Help You to Succeed In Your Search for Him

Words by Archangel Michael
20th July 2011

Valuable Assets of Living a Thoughtful Life

Valuable assets of living a thoughtful life are love and helping others not believing in what you believe, but involving yourself and others in searching out the wonders of living as you do on the earth's beautiful plain.

You are beyond doubt the most cleverly made of all wonders of the universe. What does it take to piece together each little part of a human body and all that it contains? It takes brilliance beyond anything you can compare. Each one of you is given this perfect body to house your soul for your stay on earth and for you to take care of throughout your life until you have no further use.

Each approach to living life must be from the word go the computer you have as a mind that instructs every movement of the body. Without this computer, you could neither walk run nor sit, you would not see, nor hear, without it you would not live. It is taken so much for granted how you think and how you conduct yourselves on this wonderful world of earth.

You are asked just to think, reacting to the thought's persuading you as a human being from this computer how to live as you want to live. Happiness, unhappiness, love, hatred all feelings including dread are realised just by opening your minds for thought as originally was meant to be.

Your brains, your minds calculate what and how you are able to use the very beginning of this computer. Your mind, this computer, is such a brilliant work given by your Creator. It has so much room to take in and remember knowledge, it is you yourselves that cannot learn quickly and easily enough to even start to fill a cell in your brains.

Just think of how much you as a brain have considered knowledge from the beginning of time. How much have you learned, how much information has your brain retained as the universe calls.

To train your thoughts to span the future to scan what you have been given and to scan the futuristic times to extract information to fuel your ego's, to fuel your lives giving a reason to keep putting your minds to future improvements, to wondrous beginnings of furthering your minds to encouraging your brains to expand into future thinking, helping to solve disease and illnesses of body and mind all enabled by a generous and brilliant Lord and Creator.

Glory Be To God

20th November 2011

Voices of a Different Tongue

Generally when we speak in voices of a different tongue to those souls throughout your land, we find it most interesting how the words differ country to country discussing the thoughts and reasons why God your Father exists and should He exist, what does God look like? Is He white, black, yellow, red or maybe a little of each? Why they may ask is He always portrayed as white?

Your Father in Heaven is none of these colours; they just exist on the earth plain. In the minds of all races God can be seen as a likeness to themselves, why would it not be so? Why would it be any other way?

Written in this Book, a question is asked and a reply given, "**What is God Like?**" God is immense, He is ever present everywhere. He is extreme energy, a shimmering source of energy that you on earth must only dream about because it is impossible to be near Him. It is only by gaining such knowledge rising through the levels of learning, raising your awareness, can a soul gain access to levels nearer to God, your Creator.

This will take time infinitum to gain the knowledge to return to your Father from whence you came, starting your journey as a small particle of energy released from Him to form another soul to set a further life in motion, another child born from Him. It is this circle of complete love and acquiring much knowledge as an innocent, then returning after the knowledge that would allow you to be invited back to join your Father with perfect purity and extreme love with your Creator and Father being so proud of you individually.

It is where in the world that the different tongues are spoken, because God is the same everywhere, wherever the prayer originates God hears all of His children throughout the world whatever the language spoken.

There is only one God, a loving peaceful kind Father to His children, welcoming them back home with praise whosoever have lived their lives on the earth plain with God and His son Jesus in mind, living their lives with the utmost sincerity living the trials of life with genuine effort and understanding of how you are expected to carry out your life.

For those souls who have chosen to live their lives anywhere throughout the world, for them to know that wherever they are their God, their Guide, relatives and loved ones who have already passed to Spirit can and will communicate to them in their native language. Whatever their race or creed they are one of God's children and are necessary to be there in order to help run that part of the world where they have laid their roots down in order to live a life on the earth plain.

God Is With You Always

Words by Archangel Michael
14th July 2011

Words On Behalf Of a Powerful Force

To write these words on behalf of a powerful force that has given a permission to give information on a particular subject, or to give information on a Being or soul waiting with anticipated excitement at a result or ending.

We do not often participate with a connection on request from our friends on earth, so when asked we are so excited to give our soul searching ideas to write something of interest or at the very least words of great comfort to those suffering or searching for words not heard of on earth. These words giving you all an insight into our lives helping you to know what will be waiting for you when your time is right and you pass from one world into another with ease, experiencing such relief, feeling perfect contentment as you meet the souls you have known over many lives, loved ones, friends, those from the animal kingdom, it is such a perfect reunion as you cross over to your wondrous home.

Before you continue with your life here in the Spirit World you will re-enact the lifetime you have experienced reviewing all the accomplishments, re-evaluating a life's plan being surrounded by Wise Ones with extreme knowledge putting into your own book of life your lifetimes adventures, the living of a story enacted as though on a stage.

All life's enactments are kept in archives ready for discussion as to whether you coped with the life plan offered to you before your journey to earth. It is a meeting of interest and loving communication, it is a meeting of nervous reaction as a lifetime of mixed positivity, or negativity is re-enacted, brought out into the open for you to see clearly that which is a difficult time on the earth plain.

It is for some on earth there dearest wish to be included in a book allowing their name to be written into history, for their name to be read by many allowing them to go down in history remembered by all who read the manuscript.

Each and every soul that live throughout the earth plain and spirit world have their own book of life with every moment of their life's experiences recorded for that knowledge to be recounted at any time by those powerful Beings who watch over lives with extreme love and interest.

It is only possible for each soul to know at what stage of their earthly life they may have strayed from the true pathway, by having their life re-enacted for them to experience over again in an actual being there situation, to know and feel the sensations that go along with the discovery of a life's situation whether it is good or bad, to feel the pain inflicted on another whether of the mind or of the body, or to have that pain inflicted on yourself. What will be your reaction after experiencing your life's course?

Living your life loving others, caring for others, helping others in need is the way forward for all, just to learn humility giving a helping hand however small that hand may be, let the strong help the weak. Show that passion for all to see throughout your world.

God Help You All With His Love

Words from an unknown voice in the World of Spirit
26th September 2011

To Substantiate the Greatness of Truth

For me to enlist your help for such an important venture is a word of trust in you to help substantiate the greatness of truth of words given in trust and love of one nation to another.

Jesus is your work of art in all things on earth all that is human amid the teachings of all things Spiritual and Holy. The wonders of your teachings through the words given, then sending these words to believers or non-believers who are unsure of their beliefs, is an achievement well done by all those who concerned themselves with the wonders of our spiritual world and home to you all, whether spiritually aware, a believer, or not.

How wonderful to believe in all that is meant to give you a reason to believe, a reason to not fear leaving your existence on your world, on the earth plane you call home, however temporary that may be. How wonderful to look forward to coming home once your existence in life has travelled its course and it is time for your frail body to end its tribulations and return home.

We here in Spirit do all we can to enlighten you of your life's course enabling you to be free of that fear of death. We here are only able to give you so much help and assistance encouraging you to know of the existence of your Holy Father, your wondrous God. It is up to you to follow through with these teachings expanding your knowledge, so imperative in the teachings that so enlightens your lives giving you hope, love and that spiritual enlightenment that we truly wish upon you, helping you to know all that is given to be true.

Sent To You In Profound Bliss

Archangel Michael
7th January 2012

His Word Being Complete Law

It is with deep respect and greater awareness that I contact the earth world's people with my Angelic light, Holy and Divine greatness, on behalf of a high Being from a Godly brilliance and Majestic Realm.

His word being complete law, these words sent by God are to send to you all His blessing and love, delivering beautiful power and strength of energy to show to you His love and Blessings helping you along your spiritual pathway, taking you on your journey through life as was planned in the beginning.

God loves you, His awesome light of Divine energy shines above your world encouraging you to see love and kindness in all directions along your pathways of living. You are encouraged to live your lives in His image.

This blessed message is beyond your thinking but take it to heart along with your level of spiritual awareness, shining your light for Him to see and recognise its brilliance all with God's wonders of absolute splendour.

Archangel Michael
2nd October 2011

A LOVING PRAYER

Heaven is a place of love and Divine reflections.

Addressing a congregation is a lonely but loving lesson in giving our thoughts of love, reflections of our lives and future accomplishments enough to start a healing process from words and prayers delivered by me today.

God is love and his words will not do anything to offend you.

The Spirit World and Angels guard our wonderful Lord and this is why all respect and rapport is so needed to show our devotion to our Creator.

Hallelujah and reverence we sing and have trust in all we learn and hear from Spirit's words.

All is Beautiful

Words by Archangel Michael
12th December 2010

Our Universe

~✳~

Forever and Eternity

It is forever that radiates the feelings of truth as our Father necessitates the great thinking, asking of you how forever feels. Know that your home after your life's work is done is just waiting for you, forever, however long the experiences take, you do not have to worry that time has run out for you, eternity exists.

The Spirit World has been there forever and will continue forever, allowing the capacity to learn and experience all and after gaining wisdom as you progress along your pathway, allowing access to the Higher Realms of our world and home in Spirit as your learning expands throughout time, showing God the reason for allowing a closer proximity to Himself which could take an eternity, but we have an eternity. We are discussing the word forever, which is an eternity; we here in the Spirit World, your home, have forever.

How can an eternity be cut short, how can forever be cut short? This is not possible, it cannot happen because God our Father is forever and eternal. How frightening would it be to think that our world, our home in the Spirit World was going to end on a given day sometime in our future? Why should it, it is there forever.

The Universe

Universe is as the name suggests, universal, never ending, meaning more than our minds can register and take in. Within the universe are many, many planets as earth, able to house life, give breath, give a natural way to use plants for food to sustain life.

It is so natural for those souls in our Heaven, in our Spirit World, to learn by visiting other worlds taking in the environment surrounding those souls who are sent to live a life there. Some lives you are asked to live are under water, some to live on an animal orientated state, some pure vegetable, such as beautiful trees, bushes, plants, flowers. There are many more. All these adventures are an easier choice.

All who are asked to take a life on earth know beyond doubt that here it is the most difficult and exacting course to take, but the most valuable and knowledgeable. Travelling to these far off worlds will seem daunting to mere earthlings. The journey takes no time at all from Spirit World to these other worlds.

You are probably aware that we do not talk as on earth out loud but speak through our minds, speaking with each other mentally, even contacting each other via our minds to far off places and worlds in the universe.

You are asked to converse with the universe knowing that every noise that is made, every word that is spoken goes out into the atmosphere and eventually into the universe. The universe is God; He knows everything about us, how we live, how we think, what we wish for ourselves and other human beings, but also what we wish for our planet earth. He watches what we do to ruin our beautiful planet, but most of all He celebrates each time we do something good to give life and hope to all life residing there.

Help Save Our Planet by Loving Its Beauty and Loving All In Nature

Words by Archangel Michael
26th March 2011

Worlds Beyond Our Universe

Worlds beyond our universe are wondrous homes housing life different to how you on earth know it. It is impossible now for you to reach these worlds they are beyond your comprehension of distance needed to travel to these worlds.

Thought travel is our way of going forward, of entering the thought waves giving us the only way of travelling these vast distances. You talk of the speed of light; even this cannot get you to this destination in your lifetime or indeed many lifetimes, even thousands of your earth years' is not enough of your earth time to reach a destination with this magnitude of space to be conquered.

The cleverest minds that you have on earth cannot think of speeds of travel needed to enable you to reach these wondrous of worlds. The universe never ends but continues adinfinitum; the many worlds amongst the stars are indeed your neighbours they also are searching for other life on other worlds just as those on earth. Those who are far advanced than most have the knowledge to travel the universe visiting worlds of their choice.

Never has it been known before now how to beat the frustrating battle of sound waves and light waves enabling movement so exact and profound that a particle can be carried through time, through life itself, sending that particle through space at a rate of magnificent acceleration not classed as speed after reaching and conquering that magnitude of vastness so needed to reach the inhabitants of far off worlds, these thoughts are too far from your earthly reckoning.

To have in your minds "what would these inhabitants of other worlds look like or would they want to know us earthlings as friends? As you go forward into the distant future increasing your knowledge and in time obliterating illness and disease, gaining the knowledge enabling you to travel the universe at will, could put the words "mixed race" into a different perspective. How these brilliant thoughts would play an active part in your imaginations that are beyond your comprehension.

12ᵗʰ November 2011

Parables

~✳~

Simple Story Told

A parable is worth its weight in gold telling all soul's about the good that is done in such a simple way, an understanding way for us all to take notice of, enriching all who read and listen to the simple story told. There is always a happy ending to a parable. It is to teach you all how to do a kindness for another person without expecting reward in return.

Love for another is a wonderful thing to behold, to love that person or animal, giving them help through your love, is an event worth thinking of and keeping forever.

God gave Jesus these parables to tell the world because such a simple story is easily understood, then remembered over a long period of time, whereas, something complicated is not easily remembered, in fact, forgotten before the ending of the story.

One of the favourite but important parts of any parable is carried out by a donkey, a very special animal made famous of course by bearing dear Mary to Jerusalem for the birth of Jesus.

The parables told by Jesus can include a donkey and can make you think of the desert, heat, dust, and sand, it is so easy to imagine the weariness of man and beast experienced by travellers, desperately in need of rest, water and food, travelling a great distance by foot, just relying on a faithful animal to carry their load without complaint. If it weren't for the kindness of a neighbourly passer-by, the stories would certainly have a different ending.

Jesus told us many parables, they were to make us all think of what is possible, that just the tiniest, kindest thought for others who are struggling in the same world, is such a wonderful deliberation for any one of us to take in, to encourage us and continue with our lives in God's name.

25th June 2011

A Parable of a Farmer and His Family

Once upon a time a farmer with his animals needed to trek a long distance to a far off land in order to feed his family and animals, but most of all to find lifesaving water. He was given the strength to follow his own visions as his life's possessions gathered over many years were carried steadfastly by donkey and camel.

The farmer's wife would feed her husband and children on the meagre supplies carried with them from their poor home. The task to find food and water in such extreme conditions was more than could be borne at times.

All near to starving and thirsting, all they had left to help them they thought was to drop to their knees and pray to God our Father in Heaven for His help with obtaining sustenance for them and their animals who had served them well, carrying a heavy burden along the uneven rocky roads in the heat, wearily travelling to their new home, if only they had the strength to carry on without water and food.

They travelled through the desert hiding from the extreme heat, resting as often as they could with no shade to be found only from their meagre blankets and clothes that took away some of the intense sun.

Almost at the end of their strength and dying for the lack of water, they were in a desperate condition. Low and behold God came to their rescue, their aid, in front of them in the desert appeared a wondrous sight, an oasis, trees, shade, wondrous of all, water. Gratefully, they thanked Him with all their hearts for saving their lives, answering their request to be guided to food and water.

Other people and their animals were also sharing the life giving water that in turn encouraged an abundance of food to grow. The oasis welcomed many travellers in difficulty along their journey through the arid dessert. When the farmer's baskets were filled with food, casks filled with water, newly energised, they set out on the last stage of their journey heading for the new home dreamed of for many years.

At the end of their wearisome journeying through the dessert they could see sand turning to greenery as they walked. They had found home, a gentle stream of pure fresh cold water ran below the site of their new home which would help the farmer to grow all the food and water needed to live a rich life.

It had taken them many months to reach their destination, so many difficulties overcome along the way. Only prayers asking help from God enabled them all to carry on in the future with contented happy lives.

You are encouraged to ask for help knowing that with believing in your prayer your request will be heard, help will be at hand lightening your worries, lightening your load from your shoulders. Ask with belief in Him. He is love beyond all else, is your Lord and Master, your God in Heaven.

God Be With You

Words by Archangel Michael
19th May 2011

A Parable for Us All

For a few days over Easter I had not been well, therefore I had not asked for any input to add to the pages of this book. As I was feeling a little better on the Sunday I sat down and asked to receive further words, unnecessarily apologising to Archangel Michael for not lighting a candle and asking Him for more beautiful words or prayers earlier.

I explained to Him that I had not been well, that I still was not 100 per cent and that I had not thought of a subject to ask for His help with, that it would be great if he would just give me a story of His choosing, the words a **parable** were given to me and to my delight the words just flowed.

Parable

A man was walking along a dusty road with blistered feet longing to find a stream with cool water to stand his poor feet to cool the blisters causing him pain, particularly the heat from walking through the dessert with his faithful donkey carrying what few possessions he had accumulated through his life.

Continuing on slowly he passed a short train of people with camels walking in the other direction. It was with so much kindness that the Master of the train could see his discomfort and immediately offered assistance by bringing a bowl of cool water for him to immerse his painful swollen feet. Most importantly he gave a drink of cool water to both him and his donkey that was so very welcome.

What kindness that was given without asking or without pressure of any kind. After he was refreshed he thanked the Master of the train and continued on his way.

It was months later that travelling along the same road but going in the opposite direction a wondrous occurrence happened; this was to enable our friend and his donkey to repay a great kindness. The train of camels with their owners were also going in the opposite direction. They had fallen on hard times and were struggling without food or water.

It was such a wonderful thing to see a poor man his only possession his donkey, laden with his food supplies to last him for many months. All was spread out on the side of the road filling the bellies of the camel train and giving them water to drink. It was not as though there was an excessive amount of food on the donkey, but when shared with those without, there was plenty enough to feed those who were hungry and thirsting.

To share all that you have, to show the love for one another increases our chances to live a kindly, Godly, loving life, showing to others how this is done.

Words by Archangel Michael
Delivered To Us by Arlinda
24th April 2011

ANOINTED PRAYER

Let it be known today that this prayer has been anointed by our Father Himself as a Blessed encounter with both Him and You all on earth, who are in need of a wonderful and generous meeting of minds with our dear sweet Lord.

Dear sweet God we ask your help to resolve wars, we ask you to resolve the cruelty residing in our world that we are not able to deal with on our own.

We ask for your help to teach us all how to live together and to love each other as asked of us in your name.

We ask you to take away the cruelty of all animals letting them lead a pain free existence alongside us humans sharing our life's path and we helping by sharing theirs.

Help dear Father those starting out in life to be loved beyond all else, so that their childhood memories give them a strong, happy and contented start to their life's purpose.

Dear Father, bring to us your thoughts for us to follow, enabling us to be so proud when we follow through with your ideas and instructions helping us to live better lives.

Steer us away from the evil ways of others not yet following in your beautiful aura, we look to your light for guidance with everything making us richer for your extreme beauty, strength and Holiness.

We ask for your love always. Keep us in your sight through thine eyes forever and for always.

You are indeed our Divine Creator and we ask you to help us keep our world a Glorious place to reside living alongside the trees, forests, flora and forna rejoicing when this teaching is known throughout the world.

Help us dear Father surrounding us with love and looking upon us with favour surrounding us with the light of your Divine existence. **Amen**

Words by Archangel Michael

7th June 2011

Poems

~✳~

Loved One

Doth the Wild West wind sadden when
The summer breeze takes its glory?
Doth the sun feel lonely when
The moon replaces it in the midnight sky?
Doth the sea weep when the earth
Turns and it must leave the shores?

Though my final leaf has fallen
I will someday return like the Wild West wind
Don't shed a tear for me now, I feel no sadness
The sun will return tomorrow, the sea will greet the shores

As I watch the earth turn, I feel peace as I am the sun
No need to shed a tear, Loved One
I am at peace, the moon and the tides now

Freedom greets me at these Golden Gates
And I will follow you, Loved One
Maybe not this lifetime, but I will greet you, someday
Until then, Loved One

By Lucy Phillips
Aged 12 years
A dear granddaughter

Written with Love for her grandfather to remember his son who passed to his home in the Spirit World 28th September 2008

Happiness

As the ripples of waves
Lap the shores
The sun hides behind
The Horizon
My little rowing boat did bob

T'was my soul that did sail
Upon my boat
My happiness that did
Keep it afloat
My little rowing boat did bob

Until one stormy night
A tide that came
Stole my little rowing boat
Away from me

I sat and waited every night
To see my soul
That did once sail upon
My little rowing boat

One cold winter's morning
I found my soul
Clutching a piece of rotten wood
From my boat

Upon it was a simple word
'Happiness' it said
My friend had returned from the awful storm
My little rowing boat did not

By Lucy Phillips
Aged 11 years
A much loved granddaughter

A Child's Prayer
By Angel Pippa Rose

Little children are we
As delicate as a rose petal
So perfect and so pure

Send dear Jesus your softness of love
While your light shines around us all
We ask dear Jesus that you keep us safe
Staying with us through our sleeping state

Love and cherish us in all we do
We are not always perfect it is true
But in thine eyes dear Jesus
See us as your little angels of surprise
Holding us in your loving arms of kindness
Smiling down on us with your loving eyes

Love all little children sweet Jesus
Let us see you in our dreams
Let us hold your hand while you take us
On a journey of love and surprise

Love us dear Jesus
Let us know you are there
We know that you love us
That you really do care

We love you so sweetly
Dear Jesus with love
We look to you in Heaven
Looking down from above

11th July 2011

Carole M. Day

I was finding it quite difficult to receive this lovely prayer. I could easily have stopped trying and continued with other words but I felt that would not be fair, that maybe the sender was having difficulty sending to me, as it turns out the sender was a young Angel Pippa Rose.

The disappointment might have been hard to bear for her not to be successful sending this beautiful prayer. Having already been told that there were many in the Spirit World wishing to be included in this Book I persevered.

A Childs Prayer, thankfully became very clear, how delighted I am to add it to the book. After receiving the prayer I asked who had sent it to be included and this was my answer.

"I am an Angel and my name is Pippa Rose. I am a small Angel as in your world just a child. It is with excitement that I ask for my prayer to be included in the Book. I send to all those children who will read this loving prayer, my caring love and happiness for all those dear little children in Jesus 'care."

Jesus

~✳~

All This in Jesus' Name

Jesus' name is known throughout Jerusalem spreading further afield to worldwide. He must be known to all souls as the Son of our dear Father in Heaven.

Jesus is kind, long suffering Holy of Holy's giving His all for the sake of all on the earth plain. He is so Holy, so loved by all in Spirit and it is such an event to meet Him.

His aura, His Holy beauty is all around to share with all who ask of Him. Our Father and Creator in Heaven sent His beloved son for us all to see. Stand not aside but ask for His presence and help. He is our complete Saviour who will comfort us all and also teach us what is expected of us, to help us see and reach the light.

Have you asked of our Lord Jesus and Saviour to save our world? Have you asked Him to help little children see His light and hold His hand? Jesus loves us all but children are in His care always.

He said 'suffer little children to come unto me', however their lives are lived, whether in a happy contented home or some cruelly, at the end of their life's stay they are welcomed into the arms of Jesus.

All This in His Name

Words by Archangel Michael
16th February 2011

A PRAYER ABOUT JESUS

Divine Father and Creator of us all Creator of our worlds, spirit or earth world help us in our time of need.

Hear our prayer; relieve the suffering bringing light to their lives, lifting their thoughts into the light of our Realm.

Jesus is a Heavenly Body; his soul is beyond Holiness as you know it. He is there waiting for your thoughts. He is joyous and delights in hearing from you.

For those of you who have already seen the light, His light, He sings His praises. Each new speck of light that reaches Him is a wondrous revelation.

Jesus is surrounded by his children; He is surrounded by the animal kingdom. He is in Heaven, He is Heaven.

Awaken your thoughts, your lives and spirit to finding Him, surrounding him with your love.

God Be Praised

Words by Archangel Michael
16th February 2011

Jesus Christ Our Savour

Dear Jesus Christ our Saviour who suffered as his earthly life was taken away. As we all know who believe, it was His destiny to suffer on the Cross as was instructed by our Dear Lord and Master, His Father in Heaven.

He suffered for us, for us to see and experience and know how He, as a beautiful spirit, a Divine soul, was shown and given to us so we will remember all He was and did, healing, miracles, caring for the needy, those ill and suffering, those who asked His help, those who half believed or didn't believe at all.

To see and to feel Jesus about you is a beautiful feeling. He is all inspiring, all complete love, a magnificent energy that is beyond our knowing. Jesus is kindness, all loving. He is just perfect and to believe in that perfection and ask Him for help is all that He has worked for and hoped for from us soul's here on earth.

Jesus is energy, pure energy, a brilliant light that illuminates around Him, a Godly light that He is willing to share with you if you ask.

Trust in our Father do not err
Keep his revelations deep inn your heart
His love is all around
Why would we cry?
Happiness is forever if you wish it thine
Talk to Him, sing all the while
Say how much you love Him
His love is purposely thine

Jesus Is Waiting

Jesus is waiting to hear your proclamations of love to His sweet Father, your love sent with prayers acclaiming His Glory and Holiness, telling Him of your exquisite love sent with happiness, looking to His light of brilliant perfection.

Joy to you all on earth, we send to you the love of our Lord Jesus who watches over you with His light of vision, a smile of love and kindness, a vision of hope for you wanting your lives to be lived with fulfilment and encouragement to love one another, which in turn gives you the love to pass on to another life so in need of rescue.

Sanctify all those prayers and Holy thoughts with meaningful approach giving hope to all those dear souls in need. God bless all those souls who live spiritual lives whose beliefs are manifesting into a much enlightened approach, giving our dear Lord much to applaud and be joyous.

These developing souls are more able to spread the word of their enlightenment, showing each small step taken to further their earthly development, progressing along their spiritual pathway.

Jesus' life on this earth was to show us the way to His Father, to our God and Creator suffering much by the hands of non-believers. His spiritual stories live on through time for us all to share and to take with us through our lives.

Now is the time to memorise the Divine stories told in the form of a parable or a story told in the Lords bible. Since childhood memories served you well, remembering those things taught to you about Jesus, Mary and Joseph, the Archangels and Angels, beautiful pictures conjured up in your minds of love and kindness, childhood memories of loving things to be remembered putting you in good stead for future thoughts, to believe or not to believe.

To all those who were not given that wonderful start in life of loving memories of stories and parables we send to you our Blessings and God's light and love to help you on your spiritual pathway.

All countries throughout your world have places of worship, different ways in which to worship the Creator your God. Stories are told through the ancient stained glass windows in delightful colour and delicate design throughout Holy places, Gods houses of worship.

If you open your eyes to the learning signals that are given throughout the glorious stages of your lives that are throughout the world for all to see and notice. To stand back and take notice of all the messages sent from the Realms of Spirit, the Realms of Light.

It is for you to recognise the signal sent for you to notice, to be excited by the very thought of a home existing where you will return when your lifetime here on earth has finished. We wish for you in your existence on earth to believe above all else those stories, those parables told by our dear Jesus, giving you that insight into your beautiful home that is just waiting for your return to the World of Spirit.

God Be With You

Words by Archangel Michael
12th September 2011

Knowledge Of Our Dear Lord Jesus

The knowledge of our dear Lord Jesus is a fascinating area in which to dwell upon. Who on this earth plain is knowledgeable enough to feel they can enlighten us as to how He felt, how He suffered just for the people on earth.

How brave was He to withstand knowing He was sent to earth to suffer for us all. How strong must the love have been for Him to be aware of what was in store for Him, just to help us here in the name of His Father?

Jesus talked of love all the time. The love we think we know we feel which is so weak in comparison to Him and all the Realms of Spirit. It isn't until you do pass to Spirit, to Heaven, that you realise what strength love really is.

God loves us all so very much that He sent His only son to help us on earth to follow a godly and loving life. To awaken those souls who are not sure what it is all about, who He was and who His Father is. What a wonderful place to desire to be, this is where you will be.

It is only by looking into books, reading experiences that can help you proceed in your journey. You will pass over to the Spirit World which is called your home. In the meanwhile, why not enjoy the journey we are talking about, that you are being taught, before the time is here when it is your time to go home.

Heaven, the Spirit World is light. We are asked to look to the light, imagine all being light and not darkness. The higher the Realms you travel to the more energy the more light you experience, realising the more you are unable to travel further because of the strength of this pure energy.

It is only by knowledge you amass as you go along your life's journey the more you realise God is there and not an unbelievable story, a figment of imagination. The more then do you become enlightened taking you to higher plains where you are able to withstand more light and energy as you go. God is the highest Realm, we are not able to imagine such beauty, splendour, love, energy and light, and we are not able to go to such imaginable lengths.

We souls who are so loved were born of our dear Father, nurtured, loved and encouraged to grow, whilst gaining experience enabling them to live as a soul in a body on earth. The journey is to gain access to higher levels or realms; the more we learn and believe eventually being nearer to our Father over much time.

Remember we do not have time in Heaven, time is adinfinitum, forever. Just head for love, head for light. You are being taught now in a small way, how much you believe or take in is very much up to you, read and digest all information given to you. Also remember, this information written in these pages are sent to you by a Divine Being and are written in Truth.

The Angelic Realm is beauty above all else, colour surrounding Glorious Angels. The word Angelic means soft, perfect, Holy beyond all else and gentleness. We will pick out the word perfect because God made them just that, perfect. Love beyond imagination lies in the Angelic Realms. It is difficult to know and understand how hard these Beings work for our salvation. Cherubim and Seraphim are other Divine Beings serving God, whilst protecting Him.

A GLORIOUS PRAYER

Father in Heaven, Holy Ghost, Holy Spirit honour thy helper with gratitude, love and honour.

Send our thoughts, love of humanity, to all who believe and are willing to succeed in learning about our Home, Heaven, and Spirit World.

Gratefulness is needed to encourage brilliance in light and endeavour to succeed in your learning's.

Be strong in your learning's. Listen to the stories of the Lord. Believe and sanctify the Cross learning to extreme belief in yourself, Himself, and all spiritual learning's. God loves you all.

Far from believing the opposite of religion on all sides of religion, love God with all your hearts and believe that He is with you always whatever the circumstance. Believe in Him who is Creator of all.

He will welcome you home when the time is right. Head for this occasion with all your might: because it means everything to Him in mind and body.

Love and listen to His voice always believing in his word and what it means to us all.

God Bless You and Keep You in His Sight Always

Words by Archangel Michael
17th December 2010

About The Author

~✳~

About Me, Carole

It is with wonder that I approach this book as throughout my life I would not have dreamed it possible. At the age of 68 years I am attempting what I thought impossible, but with my Guardian Angel, Archangel Michael by my side, the precious words and prayers for this book just flow into my mind and with pencil at the ready, are written down for posterity to help those who wish to read God's words, straight from an Archangel, His Divine helper.

When I was a child I attended St. Michael and All Angel's school and also attended St. Michael and All Angel's Church at Sunninghill, Berkshire. It was not until later years' that I put two and two together; there must be some link to me having a Guardian Angel, Archangel Michael.

My first major spiritual experience that I keep in my heart always and remember more than any other was when a beautiful Being, a man dressed completely in a sparkling white suit was standing at the end of my bed. This experience did not frighten me, quite the opposite, I remember distinctly saying "hello, how lovely to see you, what are you doing here?" He smiled at me I remember, walked around the bedroom, out into the hall and down the stairs. Immediately, I grabbed my dressing gown from behind the door and rushed downstairs after this beautiful Being, but alas he had gone. It was so obvious to me that I knew him and recognised him; it was also obvious that I was not meant to remember his face or who he was.

You have probably read other books, but after reading this one, branch out and read many more. It is only by reading others' experiences that you may say, I too have had things happen to me but have in the past put no store by them. Well here you are thinking of your happenings and starting your spiritual journey, how beautiful is that, what a privilege?

I was given along my spiritual pathway many experiences that would fill many pages in themselves. It was only following family to Cornwall where my husband Chris and I set up home in Truro in 2008, that it became evident to me, that spiritually this was where I was meant to be. This was a good move for us but mainly for me, finding a teacher from the local newspaper to guide me through thoughts and experiences and finding reasons why.

The first contact after the initial chat found me arriving at a small village hall joining fourteen other hopeful, interested, eager pupils. We were all greeted by the teacher, our names were taken and a very small piece of paper handed to each of us with the explanation, "written on this paper is the name of your Guide".

The name Michael was written on mine and I thought great, that's a nice strong name. To my surprise later in the evening the name was extended to Archangel Michael with instruction that I must have respect always and use His proper title. This information gave me much to think about, so much doubt that such a wonderful Being could be assigned to me. Yes, doubt, maybe the teacher was mistaken with the name I was given. Perhaps it was George, John, Margaret, Peter; the lack of confidence in me was evident. Over time I have had the proof needed to show He is my Guardian Angel and Guide, the feeling of awe and wonderment will be with me always.

The course at the village hall ended, much to my disappointment, I had enjoyed the experience very much I didn't want the journey to end. What do I do now I thought; the idea struck me to ask the teacher if she would be willing to give me tuition weekly on a one to one basis. That is exactly what happened, my lessons began.

An exciting stage on my spiritual journey was seeing the colours of purple and gold, each time I relaxed or meditated. The colours were so bright swirling, mixing with each other, and just watching from within, eyes closed, just enjoying the wonders of Archangel Michael's colours, it got to the stage where I would just ask to see. For about 18 months I experienced this wonderful occurrence and then without warning, was gone.

It worried me that He had gone from me, I wondered what I had done, or said to make Him go. Now I understand that you are given so much to experience along the way before you are asked to move on to another.

The aim was to become a Medium, this did not come naturally to me, but each lesson I had, brought me closer to that end. My teacher encouraged me to go home, meditate and ask my Guardian Angel to send a prayer or words through me, this I did. What a wondrous, awesome thing to happen to me, to experience. The words and prayers in this book were sent through me for all here on the earth plain to read. I feel so thankful to have been chosen.

I am beginning to wonder as I write if my stories are of interest to those who are reading these pages. How difficult it is to put words down on paper to keep the readers interest and for them to be waiting for the next happening in the chain of events of my experiences. It is with love of my thoughts that this expression of interesting stories continue along the way, hoping that you still follow.

Continuing my lessons and meditating, I was asked "what can you see?" The eye of a horse was clearly in my vision and was for several months. Not being sure what the meaning was, I could only assume that as I worked with horses and loved to ride in my youth, it was second nature for me to be involved spiritually with this wonderful animal and maybe it meant clear vision.

One meditation when asked to follow along a path, through flowers, through woodland, arriving at a beach with yellow sand and blue sea, we were asked to stay there until told to return. It was such a lovely warm day, in the distance I could see a horse galloping toward me, white, tail flying, I just looked in wonder as it slowed and came up to me snorting and whinnying. After a while I asked his name, sensing his reply "SALT". We are told never to doubt what we are given, so Salt has been my friend ever since that day. At the start his name seemed too ordinary for such a wonderful animal; it became spiritual and beautiful, belonging only to him.

Another meditation, Salt galloped along the beach towards me a Red Indian chief mounted on his back. He had a magnificent feathered headdress, feathers of all colours reaching down to his feet. It was obvious they were a team. Even with the wonderment of it all, I was pleased with myself that I asked his name, he replied "I am Running Cloud of the Navaho Nation."

An everlasting story of a tale, a prophecy to be read from a manuscript, is a thought provoking idea. We all believe in our own experiences, looking back at our memories, finding them so thought provoking, so real, simply because it is an experience that you can believe in, because it has happened to you.

Imagining another mind taking in that story, thinking do I believe that which I am being told. It is your thought, your experience that you know you must share with others, so they can also pass on that happening.

If you find you are in that position of writing a thought, a real happening that is real to you, just have patience, just stay strong in your belief that passing on to another soul can help them to learn with the teaching you have unfalteringly passed on to enlighten a further soul.

It is so easy to say I don't believe in what I am being told. How wonderful it would be to believe all we are told. A mind, a brain is there to take in data, information to be taken in and digested, allowing you to take out the bits you can relate to first, then fathom out those ideas that you are not sure of, arriving at a solution you can pass on to others with pride. Delve into your hearts for love and answers to your questions and trust.

The title Snatches of Brilliance is a normal idea for a name of a spiritual book. It is thought that maybe there is another book with a similar title. It is the word God that has been seen and used for many excellent stories, each explaining in themselves or trying to deliver an explanation of what Heaven or Spirit World is like. Think about this, if you have not been there how could you know unless a soul already there can talk you through with ease where your home is and what happens when you get there.

Stories in this book or other books cannot liken you to the real thing or experience. Now that is where your mind and brain comes in to it all.

A human beings imagination can go a long way to introducing you to the Spirit World. With only a little ammunition with which to start the process going, for example reading books as I have already mentioned previously, and talking to others'. Isn't it amazing how much you have to think about?

All the wondrous beautiful Beings of all the plains' in Spirit, from the not so knowledgeable yet, to the teachers who have learned over perhaps many earthly years' after many visits and lives. This now asks you to ask yourself, have I been here before? How spiritual am I? Am I enlightened now? If not, how far away from that Knowing am I? How much do I believe about God our Father, Jesus Christ our Saviour? Who died and suffered for us on the cross, so we can recall that day in time and thank God for Jesus, otherwise what would we have to set our rules by, who would we have to set our rules by.

All those Beings from the light who guard God, who help us all on earth, if they see a tiny light from enlightenment starting to happen from one of us, they are so delighted with a great deal of celebrating going on in Heaven.

In the earlier years' before I had any thought of enlightenment, though certainly believing in something, it was a regular occurrence for me to float to the ceiling out of my body, this they say is an out of body experience. It was worrying for me; in fact, I closed the bedroom windows just in case I floated out and away, never to be seen again. Not knowing the truth about this occurrence was a frightening thing, but having it explained about the silver cord attached to you stopping the soul from leaving the body unless it is your time to pass to spirit, then it breaks automatically allowing you to go home.

I remember at that time seeking a hypnotist to do something wonderful such as stopping me from smoking without any effort from me. Forty to fifty cigarettes were the norm for me to smoke then, so off I went to the hypnotist puffing nonstop until I arrived at his front door. I was so ready to be told you are a non-smoker. I giggle remembering visiting this man, handing over £20 and after the session thanking him very much for what he had done, rushing to the car picking up the lovely cigarette stub I had stubbed out with such enthusiasm before entering his front door, then lighting it again with as much enthusiasm puffing away as I drove home.

The most informative part of that visit was telling the hypnotist the story of floating to the ceiling which I explained I found quite frightening. He was a very interesting man knowing more about spirituality than he first let on and just talking to me helped enormously to overcome my difficulty, just simply by explaining that if I anchored myself to the earth I would not rise up to the ceiling, I did this and he was right.

With love and light
Carole

My Own Short Story

~✳~

My Own Short Story

I am in Spain on holiday at the time of writing this sitting in the sun writing a prayer or words given to me from my Guardian Angel, Archangel Michael, to add to the pages of the book. It is easy for me to write exactly what I am given to the letter, not altering a single word, just listening and writing what I am being given from such a wonderful source, such a beautiful Being. Well actually it is more scribbling than writing but so very exciting deciphering the words given to me to know what they say and mean.

My home is in Cornwall, England, a beautiful spiritual place, most of the villages or towns are named after Saints. We travel to Spain from Plymouth on the ferry to Santander, which is a restful journey having a cabin for the night. Once again, I am able to say that receiving the words and prayers whilst travelling is done in such a relaxing atmosphere, so long as it is a calm crossing of course.

It is a problem for me to put my own words down on paper as I am not a writer in my own right. Trying to make a sentence, paragraph, or even a few words make sense enough for a reader to enjoy what I have wanted to say and for it to be correctly written is difficult for me. I really feel that I should add words of my own, not relying purely on what I am being given, so I have asked for some help with this, I also ask your patience.

Perhaps I should start by explaining my surroundings, just sitting in the sun by a pool which I add is much too cold to even try and lower myself into. It is not a large garden but pretty with geraniums in pots and bougainvillea flowering up and over walls, in fact, very Spanish. There are many sparrows tweeting, each vying for a place on the feeder hanging from the branch of a fruit tree. It would be good to continue with the quiet ambience of the surroundings and those tweeting birds, but the background noise is taken up with an earth digger moving rocks, earth and whatever else it does to break the peace of the late afternoon, but then that is also Spain.

Each Wednesday morning here in Spain I attend a development circle where I and other interested people meet to be encouraged by experienced mediums on how to bring out their spiritual knowing gently and safely. I really do enjoy going for the couple of hours and feel that I learn something interesting each week. Not only that but making friends with other likeminded people, sitting with a coffee and a biscuit, just chatting amongst ourselves discussing what happened in circle and generally trying to put the world to rights.

My husband Chris and I live in the wonderful City of Truro just a short walk to the shops, doctor's surgery, in fact, all those important places that life in general throws at you for your needs. Truro Cathedral, now there is a great place to visit just to sit listening to the cathedral organ playing lovely music, mulling over spiritual thoughts as you gaze at the beautiful stained glass windows.

Surrounding the Truro area are many small churches, 15th century and some much older with their so interesting graveyards where it is difficult to see the names and dates of those who lived and were buried there because of the age of the stones. I consider myself so fortunate to experience all that I have been given; there is no doubt in my mind that our Father, our Creator exists and is there for us, also that Jesus is there for us too. Our real home in Spirit is waiting for us to return to and all our loved ones are there to greet us, so to visit these ancient churches and read what people had to say then about God, Jesus and beliefs is wonderful. Considering the dangers of belonging to a particular faith, speaking your mind, depending on who was monarch at that time, some were very brave indeed.

Before moving to Cornwall, I went to the Spiritualist Church in Camberley, Surrey and the development circle there for a short while which was so interesting, they were such lovely helpful people. That was the starting point of my spiritual journey I believe.

Since moving to Cornwall I now go to the Spiritualist Church just outside of Truro called Playing Place with my second eldest daughter Sally. It is such a lovely service with interesting Mediums giving their all each week.

Importantly, there is much laughter that lifts the vibrations in the little hall, the reason being not many of us can sing, or we sing before the music starts, or even that the music stops mid song. Anyway, the fact that it is all less than perfect makes it so much more enjoyable for us in the congregation and most important I feel sure, Spirit. There is nothing sombre about our little church.

I can remember earlier in my life, in my youth praying for anything and everything that I felt I needed, not knowing if there was anyone there listening to me, not knowing if I would receive the help I had asked for but still asking, giving it a try. It was asking for myself though and not thinking of others at the time, but then I feel sure all youngsters are the same until they can experience for themselves in later years, come to a conclusion, act on that, continue on their way with what knowledge they have collected over time. Maybe pray for someone else in need.

My childhood was a happy one; my parents were lovely kind people who tried their utmost to bring up their family as best they could. My father Harry was a builder; my mother would ride her bike to do housework in a posh area of Sunningdale, my brother Harry junior passed to Spirit in 1992 when he was 52 years of age and how we miss him to this day.

It would be great to know that you have enjoyed my own short story and are not bored silly. It really is just for you to relate to me, the person who has received the words and prayers for this book, the words I am so proud of and want to share with you all out there. Firstly though, you will need to have bought the book, I would so like to instil the love and the thoughts that have been put into words into this book, then into your minds so that you will remember all that is written and above all else remembering where the words originated from with all the love and light possible to instil into your soul.

With Love and Light - Carole
17th April 2011

My Experiences From Circle

~✳~

My Experiences from Circle

I have mentioned previously that each week I and others attend a development circle in Spain. It was also mentioned that whilst meditating I have had some wonderful experiences to put to paper for you to read and I would like to add more. Firstly, could I remind you of the words entitled the Parable that were given to me 24th April 2011 before I continue with my experience of 2 days later?

Meditating is a very peaceful, relaxing process enabling you to put aside daily worries for that short period of time. On this occasion we were led along country paths with bushes, plants, trees, colours of all descriptions, walking through fields until we reached a clearing in a wood where we took a seat, inviting our Guardian Angel or whoever else from Spirit who would be kind enough to join us there.

Just sitting, so relaxed, listening to beautiful piano music following the guided meditation, I began to think that was all I would experience this session which was fine by me as the relaxation was wonderful enough. But then, in my vision walking very slowly was an old woman wearing a long cloak that covered her completely. I asked her who she was and what she wanted and she replied, "My name is Arlinda and I have brought you a Parable of Kindness" handing me a sheet of paper as she continued walking slowly out of vision. In hindsight I would have loved to say, "I was given that Parable of Kindness two days ago, why again?" but it was not to be, she had gone.

At the same circle meeting we were asked to bring a photo of a friend or loved one who had passed to Spirit, a colleague would try and do a reading from that photo for you. The teachers at the circle were so pleased at our progress because we all did so well giving our messages; it only proved to them how much we were all developing.

The photo I was given was of a man and I sensed by holding the photo in both hands that he was an outgoing man, I could see him laughing in my mind. The names of Jack and David were given very clearly to me and with Jack still laughing I could see laid out at the bottom of my vision soft fur or perhaps velvet I was not sure although I was able to feel the texture, the word black was also given to me. We were all asked to tell what we had received from Spirit and to my great surprise all I had given was correct. Jack was the man in the photo, a grandfather, he was in the Welsh Guards wearing a black furry bearskin and this was how he was remembered. David was Jack's son also in Spirit. How wonderful was that, I felt so proud, it is so easy to doubt what you have been given. Each time you have success, it helps for the next time to trust in yourself.

I recall whilst in circle in Cornwall and meditating I could see a Being, tall, long hair and wearing a flowing blue gown and unusually of all the Being was carrying a large basket. After explaining this to my teacher she asked was the basket full or empty, all I could reply was "I don't know, I did not notice" it was so wonderful to see somebody like that at all, it all happened so quickly. Sharon my teacher came back very quickly saying that I had seen Archangel Barachiel as that Archangel always carries a basket.

The following week my daughter Sally and I went to an evening circle and of course meditated. When the meditation had finished we each were asked to say what we had experienced, Sally normally saying that she had experienced nothing but had just enjoyed the relaxation. It was my turn to be asked and I explained that I had seen Archangel Barachiel but not carrying a basket as before.

Sally was sitting on my right and when asked about her experience she said I have been asked to hand my mother a basket and she added that it was full of oranges. Oh, I was so thrilled and immediately said to Sally no wonder my Archangel was not carrying a basket; you had it Sally all the time to give to me. How wonderful. It was Sally's first experience in circle and she was just so delighted and has never forgotten that story.

Trying to think of a reason why there were oranges in Archangel Barachiel's basket at that time was difficult. We arrived in Spain shortly afterwards when the most unusual thing happened, we looked into our garden from the upstairs window and there were men sawing down our orange tree. Admittedly it was not a very successful orange tree but still it was ours. Rushing to the garden shouting and arms waving because we do not speak Spanish the men just stared at us in amazement realising what they had done. They were gardeners who should have been down the road taking out an orange tree there and not ours in our garden. Happily the tree was replaced by another that gave us masses of the most beautiful sweetest oranges that were a delight to eat. Well, I can only say how wonderful that is.

A dream that I experienced many times was of flying. At first I was in a small room practicing and flapping my arms like mad. I persevered at this for a long time finding that if I flapped hard enough I would raise myself off the floor by a few inches. The more experienced I got at it, I was able to run and throw myself flapping my arms and raising myself a couple of feet off the ground.

Eventually I was able to fly very well asking anyone who was passing if they would like to experience flying with me, if they said they would I would take hold of their hands, flap the other arm and up, up and away we would go. I laugh at this memory because it was ok going up, it was when it was time to come back down, getting as far up as you wanted to be, then looking down and that feeling of falling was horrible however much I flapped, but I would not let go of the other person's hand.

Love and Light
Carole

Loving Memories of My Family

~✻~

Loving memories of my family

My dear father passed to Spirit some years ago leaving my mother on her own. Both my husband and I were keeping her company one evening not long after he had gone.

A favourite program of them both had just started, the setting was in the local pub in the bar, the main character of the program was there and said quite out of the blue, Harry Wood, my father's name was Harry Wood. We all looked at each other in amazement as there was no reason for him to mention the name. My father had certainly been there to help my grieving mother, and it really did help. Thanks dad for that message and for helping us all to know you were there with us.

~✳~

My mother lived in a nursing home towards the end of her life. She really was a loving, kind person, the family would visit her as often as they could and would listen to the various stories she would have to tell. Towards the end of her life as we entered her room, she would wave us away telling us to go downstairs where all the family were, that there was lots of food to eat. Go and have a chat with them all she said, some of them you have never met. She proceeded to tell us their names, grandma's, granddad's all her sisters and brothers who were no longer on the earth plain. What a wonderful thing to happen for her at that time, they were all with her to help her cross over to home. So we went out of her room and passed time for a while before continuing with our visit.

~✳~

My mother throughout her life had not been out of England so it was to our astonishment that she was speaking Russian on one of our visits. She said to Chris and I would you like me to arrange for you to come riding tomorrow. What a surprise that was, I am not sure that is possible tomorrow was my reply, but tell us all about it please, where is this place, I would want a very quiet horse if I did.

She proceeded to tell us where she lived in Russia, the names of her friends who live next door and all beautifully spoken in the Russian language. Then to our surprise she told us the names of the horses that we would ride assuring me that I would be ok on this particular one. I would like very much to remember more of the language she chatted in but it was all over so quickly and very difficult to remember details.

After my mother passed to Spirit and at her memorial service, I had written a eulogy to be read to the mourners during the service and this included my mother's Russian escapade. The Vicar was fascinated at the reading because after the service he took me to one side and said what a wonderful thing to happen to your Mother, her complete family helping her cross over.

~✷~

My stepson, my husband's son passed to Spirit at the age of 39 years, he was an alcoholic. He was not strong enough mentally to cope with what life threw at him, his heart stopped beating one evening ending the torment of his life on earth. We know where Christopher is, in fact, we have had many messages from him, some saying how sorry he is to have caused such grief to his loved ones, but also saying how much he loves us all and most beautiful of all how happy he is in the Spirit World with his loved ones there. We all love him so much and miss him greatly.

~✷~

These are my loving memories, my thoughts and my experiences to remember and pass on to others who are interested. Have you had any unusual occurrences happen to you? Maybe you have just put them to one side not giving them much importance, or that maybe you would not be taken seriously when telling the story to others. Just make a note, jot down these wonderful experiences for future reference. I know I would love to hear the wondrous happenings of others, they are so important to you, to me and to others who read them.

Love and Light
Carole

Carole M. Day

THE LORDS PRAYER

Our Father who art in Heaven

Hallowed be thy name

Thy kingdom come

Thy will be done, on earth, as it is in Heaven

Give us this day our daily bread

And forgive us our trespasses

As we forgive those who trespass against us

And lead us not into temptation

But Deliver us from evil

For thine is the kingdom

The power and the glory

For ever and ever

Amen

The End

Of

SNATCHES OF

BRILLIANCE

~✷~

www.ingramcontent.com/pod-product-compliance
Lightning Source LLC
La Vergne TN
LVHW011155080426
835508LV00007B/423